# *Creating* Habitats and Homes *for Illinois Wildlife*

Debbie Scott Newman
Richard E. Warner
Phil C. Mankin

Designer: Lynn Hawkinson Smith
Managing Editor: Robert J. Reber

Illinois Department of Natural Resources
and University of Illinois

Creating
Habitats and Homes
for Illinois Wildlife

*We thank the following individuals for their technical assistance and reviews of various chapters of the book: Alicia Admiraal, Bob Bluett, Darryl Coates, Larry David, Raymond Eisbrenner, Stan Etter, Don Gardner, Tim Hickmann, Dan Holm, Martin Kemper, Bill McClain, Barry Newman, Bob Reber, Gary Rolfe, Glen Sanderson, Scott Schaeffer, Robert Szafoni, Lynn Hawkinson Smith, John Taft, and Kevin Woods. The work of copyeditor Molly Bentsen is also greatly appreciated.*

*We also thank the Illinois Habitat Fund Advisory Committee for making the initial investment in this project, and for assistance in planning.*

*Jeff Ver Steeg and John Buhnerkempe, Illinois Department of Natural Resources, provided much of the inspiration and technical support needed to bring the project to fruition.*

*We especially thank John Cole, Illinois Department of Natural Resources, for his advice and encouragement. His very thoughtful and detailed reviews of early drafts of the text greatly enhanced the final product.*



ISBN 1-883097-42-8

Funding was provided in part by the University of Illinois, Illinois Department of Natural Resources, United States Fish and Wildlife Service, Illinois Pheasant Fund, and Illinois Habitat Fund.

Printed by the authority of the State of Illinois 1066249 - 10M - 7/03

Printed in the United States of America

Metro Litho/Creative Drive
4201 West 166th Street
Oak Forest, Illinois 60452

# Contents

Foreword . . . . . . . . . . . . . . . . . . . . . . . . . . . . . . . 6

Introduction. . . . . . . . . . . . . . . . . . . . . . . . . . . . . . 8
*Why Focus on Habitat? • Why Focus on Private Lands? • How to Use This Book*

*Chapter 1*
Illinois Wildlife: A Historical Perspective . . . . . 13
*How Did We Get Where We Are Today? • Where Are We Today? • Quantity • Quality • Fragmentation • Reduced Biodiversity • Where Are We Headed? • Suggested Reading*

*Chapter 2*
Understanding the Natural World . . . . . . . . . . 25
*How the Natural World Is Organized • Understanding Ecosystems • Understanding Wildlife Populations and Communities • Understanding Plant Populations and Communities • The Bigger Picture: Ecosystems on a Landscape Level • Putting the Pieces Together • Myth or Fact? • Suggested Reading*

*Chapter 3*
Grasslands, Including Prairies. . . . . . . . . . . . . . . 47
*What Defines Grassland Habitat • Grassland Habitat Issues in Illinois • How You Can Help Grassland Wildlife • Management Considerations • Creating New Grassy Cover • Protecting and Managing Grasslands • Additional Management Tips for Grasslands • Suggested Reading*

*Chapter 4*
Woodlands and Woody Cover . . . . . . . . . . . . . 71
*What Defines Woody Habitat • Woodland Habitat Issues in Illinois • How You Can Help Woodland Wildlife • Management Considerations • Creating New Woody Habitat • Protecting and Managing Woody Cover • Additional Tips for Managing Woodlands • Suggested Reading*

*Chapter 5*
Wetlands and Other Aquatic Habitat . . . . . . . . 99
*What Defines Aquatic Habitat • Aquatic Habitat Issues in Illinois • How You Can Help Wetland Wildlife • Management Considerations • Creating New Wetland Habitat • Protecting and Managing Wetland Habitat • Additional Management Tips • Suggested Reading*

*Chapter 6*
Cropland and Other Agricultural Areas . . . . . 129
*What Defines Cropland Habitat • Cropland Habitat Issues in Illinois • How You Can Help Cropland Wildlife • Management Considerations • Agricultural-Land Management for Wildlife • Additional Wildlife Management on Agricultural Land • Suggested Reading*

*Chapter 7*
Special Features and Supplemental Practices. . 153
*Brushpiles • Rockpiles • Perches • Nest Boxes and Artificial Cavities • Nest Islands and Platforms • Old Buildings and Other Structures • Suggested Reading*

*Chapter 8*
Backyards and Other Small Tracts . . . . . . . . . 167
*Backyard Habitat Issues for Wildlife • How You Can Help Backyard Wildlife • Management Considerations • Wildlife Landscaping Needn't Be Wild • Developing a Plan • Creating and Protecting Woody Cover • Creating and Protecting Herbaceous Habitat • Creating and Protecting Aquatic Habitat • Creating Winter Food Plots and Wildlife Feeding Areas • Other Additions to Your Backyard Habitat • When Wildlife Wear Out Their Welcome • Suggested Reading*

*Chapter 9*
Planning: Your Roadmap to Successful Habitat Management. . . . . . . . . . . . . . . . . . . . . . . 189
*Obtaining and Developing a Map • Written Inventory and Historical Records • Determining and Recording Your Objectives • Developing a Plan of Action • Suggested Reading*

*Chapter 10*
Follow-Up: The Long-Term Plan of Action . . 201
*What to Expect from Your Plan • Reviewing Your Plan • Monitoring and Documenting Your Results • Magnifying Your Efforts*

Index . . . . . . . . . . . . . . . . . . . . . . . . . . . . . . . 209

# Foreword

*What is your interest in wildlife? Is it birdwatching, hunting, conservation, or maybe photography? Whatever your reasons, we're glad you've opened this book, and we feel confident that you will find new ways to appreciate and interact with the many forms of wildlife on your land or even in your own backyard.*

Wildlife do so much for us in so many ways. Viewing wildlife can give us endless hours of pleasure. Can you imagine

- a white-tailed deer with her fawn?
- a ruby-throated hummingbird feeding on columbine nectar?
- a red-tailed hawk soaring high above?

Wild game species are important recreational and economic resources. As food they give our tame palates welcomed changes in taste and variety. Can you savor

- roast duck with wild rice?
- venison loin chops smoked over red oak?
- pan-fried crappie fillets and morel mushrooms?

Wildlife fill niches in natural cycles that help nature function and support all life, including ours. They give us a feeling of connectedness to the rest of life on earth. Can you sense this connectedness in

- a fox hunting a cottontail?
- honey bees pollinating apple blossoms?
- field sparrows feeding caterpillars to their nestlings?

Given a chance, wildlife make our lives richer. But in turn, they need our help. This book will give you a very practical look at what you can do on your property to attract and benefit everything from blue-winged teal to butterflies. The goal is to give wildlife a home.

FIELD SPARROW FEEDING NESTLINGS

All good homes have a solid foundation. More and more, we are realizing that to benefit wildlife we have to consider everything from the ground up. All species depend on healthy plant communities for food and shelter. This book gives practical advice on establishing and maintaining such communities.

Additionally, we may have to think beyond our own property lines. Here is where this book may go farther than other publications you have read. The book focuses on what you can do on your own property, but at the same time it takes a broader landscape-level view, so that what you do on your property makes the best fit for your local area. It's just another example of "thinking globally and acting locally."

The first chapters help you see nature's continuity and understand how the natural world works. Such insights will help you appreciate the problems facing wildlife.

The middle chapters discuss specific habitat types and what you can do to attract wildlife to your property. Whether you are interested in waterfowl, songbirds, deer, gamebirds, or any other species, big or small, there are sections to help you.

*Chapter 3 (Grasslands)* can help you establish and maintain grassland habitat that will attract species such as

- Cottontail rabbit
- Bobwhite quail
- Eastern bluebird
- Ring-necked pheasant

*Chapter 4 (Woodlands)* gives you guidance on improving and establishing woody habitats that provide food and cover for species such as

- White-tailed deer
- Great-horned owl
- Ovenbird
- Bobcat

*Chapter 5 (Wetlands)* can help you provide wetland habitats that attract a variety of wildlife, including

- Wood duck
- Mink
- Bullfrog
- Green-backed heron

*Chapter 6 (Croplands)* can help you make farmland more habitat-friendly for species such as

- Red fox
- American kestrel
- Red-tailed hawk
- Fox squirrel

RED-TAILED HAWK

*Chapter 8 (Backyards)* can help you make areas around your home attractive to wildlife such as

- Cardinal
- American goldfinch
- Gray squirrel
- Tiger swallowtail

The last chapters help you to plan, implement, maintain, and evaluate your projects so that the habitat you have developed benefits the wildlife that share your property for the long term.

# Introduction

*Illinois: Home to twelve million people. Also home to fifty-eight other mammal species, 383 different resident and migrant birds, 104 types of reptiles and amphibians, 174 species of fish, and some 27,000 types of insects, mussels, and other invertebrates. Many Illinoisans regularly encounter the state's more common wildlife, such as the fox squirrel, the cardinal, the mallard duck, and the American toad. But a resident could spend his or her entire life in this state and never see other animals, like the elusive bobcat or the uncommon osprey. Each of these species, whether visible or secretive, common or rare, shares with us this 55,645-square-mile patch of earth that we call Illinois.*

Some people seem unaware that they share the state with any wild creatures. Most residents, however, notice and even take keen interest in the wildlife around their homes, farms, and local parks. For many citizens, no social engagement or sporting event can compare to the heart-pounding thrill of hunting and bagging the first white-tailed deer or Bobwhite quail; no nature broadcast can replace the experience of falling asleep to a symphony of frogs on a spring night or the awe of seeing a bald eagle soaring. Many people find that few artificial creations can rival the delicate beauty of a tiny hummingbird or the intricate construction of a spider's web. If we allow it, wildlife give an unparalleled inner pleasure.

Wild creatures, like the other natural resources on our planet, have intrinsic value. Humans have always seemed compelled to justify the existence of wildlife in terms of their value to us. In case their mere existence and aesthetic qualities are not enough, consider these facts:

- The Illinois economy realizes nearly $1.1 billion every year from expenditures related to watching wildlife.
- Hunters spend nearly $150 million a year pursuing game.

American Toad

- More than $550 million is spent each year on recreational fishing, and the annual retail value of commercial fishing is $4 million.
- Wildlife-oriented recreation plays an important economic role in many depressed and declining small communities.

Beyond economic impacts, there are less tangible but equally important benefits of wildlife. The vast array of vertebrate and invertebrate animal species that inhabit Illinois depend, directly or indirectly, on one another. Removing any one of those species may result in disproportionately high numbers of another. Even if we were to consider solely our own comfort, health, and ability to produce food and fiber, preserving a balanced community of wildlife is essential. Certain species prey on others that are considered to be agricultural and forest pests: red-tailed hawks and great-horned owls efficiently harvest mice and voles, which can cause significant damage in crop fields. Bluebirds and Baltimore orioles extract millions of caterpillars and other invertebrates from our croplands and forests. Other animals control the abundance of a species that is considered a physical pest to humans. One bat can consume up to 3,000 mosquitoes in a night! Wildlife are vital to the survival of the human species.

## *Why Focus on Habitat?*

Humans, like every species with which they share the earth, are inseparably tied to their habitat, or living space. But humans have the advantage of being able to alter their habitat to a greater extent than other animals. If people find that something is missing or doesn't suit them in their living environment, they create the missing element, rearrange the landscape, or eliminate the "problem."

Wild creatures, on the other hand, cannot alter their living conditions so significantly. Many cannot change their particular habitat at all. Naturalist Aldo Leopold illustrated this point in his popular book *Game Management*: "The essential difference between a deer and a man is that man builds farms, factories, and cities to provide himself with the elements of an habitable range, whereas a deer must accept the random assortment laid down by nature and modified by human action, or move elsewhere." Having a suitable living environment is essential for any species to survive. If wildlife cannot find good habitat, they must go elsewhere or cease to exist.

WHITE-TAILED DEER

In Illinois, we humans have taken nearly maximum advantage of our ability to change our surroundings—to alter our habitat. Many would argue that our actions have made life better for humans, but there is little argument about our effects on much of Illinois wildlife. While the populations of some species, such as Canada geese and red-winged blackbirds, have actually increased as a result of the human-modified environment, many other species have been negatively impacted. In fact, our landscape manipulation has pushed some wildlife, like the prairie chicken, nearly out of existence. Habitat degradation and destruction by humans has been the single biggest cause of the decline of wildlife populations in Illinois. Our ability to change habitat must be coupled with the responsibility to consider the needs of wildlife if wild creatures are to survive. And though an extinct species can never be revived, some of the damage we have done to our plant and animal communities can be undone. While we have the ability to destroy habitat, we also have the capacity to restore and protect the wildlife habitats that still exist, and to reconstruct suitable habitats on land we've rendered inhospitable for wildlife.

## *Why Focus on Private Lands?*

Local, state, and federal governments and nonprofit conservation groups own 750,000 acres of wildlife habitat as scattered parcels throughout the state. These acres do make a difference. However, 95% of Illinois' 36 million acres are privately owned—that's nineteen out of twenty acres. Here is where the greatest potential for improving habitat exists—on these private acres.

Much of the privately held land in our state is rural acreage managed by farmers. Corporations also own thousands of acres. The remainder of Illinois' private land belongs to schools, churches, small businesses—and every citizen who owns a yard or lot.

Whether you own one acre or one thousand, the decisions you make and the actions you take regarding your property affect the non-human species that reside or visit there. Every piece of unpaved land in Illinois has the potential to support some wildlife. To be responsible stewards during our tenure, we must consider wildlife in all our land-use decisions. No matter how much land you own, your ownership is temporary. How you manage that land while it is in your care will have an impact long after you are gone.

What about citizens who don't own land or who live in the city? They can still effect positive change for wildlife.

You may be on the board of a church that occupies two acres of land, neatly mowed but with little habitat for wildlife. Or you may sit on a county board that makes many land-use decisions. You may be a school-board member or a teacher at a school with idle land that could become a habitat demonstration plot, benefitting both wildlife and students. You may belong to a country club that could improve its land

CANADA GEESE AT NEST SITE

for wildlife. Even more significantly, you may work for a company or agency that owns land, perhaps even large amounts, that could be improved for wildlife.

## *How to Use This Book*

Land management decisions, large and small, are made by thousands of Illinoisans every day. With this book, we want to help citizens make informed, responsible choices and to provide practical information to turn those choices into reality on the landscape. Chapters 1 and 2 explain the foundation on which habitat management is built and should be read first. Chapter 1 provides an overview of Illinois landscape changes and the resulting effects on wildlife. It also presents major habitat issues in our state today. Chapter 2 explains the basic ecological concepts that underlie the land management practices recommended in later chapters.

Chapters 3 through 6 describe management practices for individual types of habitats—grasslands, woodlands, wetlands, and croplands. They may be read independently of each other, as needed, or as desired. Chapter 7 provides guidance on use and construction of artificial structures for wildlife. Since these structures can be used in conjunction with the management practices described in the preceding chapters, most readers will want to consider at least one project from chapter 7. Individuals who would like to do habitat projects on small plots or in backyards should read chapter 8.

The adage "If you fail to plan, you can plan to fail" certainly applies to habitat management. Chapter 9 provides assistance in forming realistic plans. Making positive land-use changes doesn't have to be a complicated effort, nor does it have to be done exactly as described in this book. But having some sort of plan will help you achieve your goals more efficiently. A plan can be as simple as a scribbled set of notes or as elaborate as detailed in this book. Even if you doubt that you'll commit your strategies to paper, we encourage you to at least skim chapter 9 to learn planning essentials and details on getting started. The book concludes with information on maintaining your habitat, in chapter 10.

A key to successful wildlife habitat establishment is access to reliable information. That is why a "Suggested Reading" list concludes each of the first nine chapters. Additionally, the Illinois Department of Natural Resources and the Illinois Natural History Survey publish a wealth of Illinois-specific information—circulars, leaflets, booklets, and the like—that has proven valuable to many Illinois landowners. Only a few of these publications are included in the "Suggested Reading" recommendations. For complete lists of publications, contact the agencies directly:

- Illinois Department of Natural Resources
  DNR Clearinghouse
  524 South Second Street
  Springfield, IL 62701-1787
  (217) 782-7498
- Illinois Natural History Survey
  INHS Distribution Center
  Natural Resources Building
  607 East Peabody Drive
  Champaign, IL 61820
  (217) 333-6880

As Illinois society becomes more and more urban, and as the pace of life and work seems to move ever faster, we are in danger of losing a basic understanding of the natural world and how it functions. It is our fervent hope that this book will remind readers of the fundamental reasons for conserving our state's environment and will inspire and guide them to take action to help preserve Illinois' natural heritage.

*Chapter 1*

# *Illinois Wildlife*

## A Historical Perspective

*Like a sculptor shaping a piece of clay, we humans have molded the Illinois landscape to take the shape we desire. And we've done more than modestly adjust its features; we've transformed it completely. Many of the underlying characteristics of the land, such as topography, still exist, but the outward appearance no longer resembles what was here 200, 100, or even fifty years ago.*

PRAIRIE CHICKEN
PREVIOUS PAGE: LEFT, WHITE-TAILED DEER; RIGHT, RED FOXES

In a mere two centuries we've turned vast grasslands into crop fields and converted forests into shopping malls and homesites. We've leveed and drained wetlands to dry them up and impounded wooded ravines to make them hold water. Meandering streams have been straightened and, in a couple of places (the Chicago and Cache rivers), we even did what previously only a great earthquake could accomplish: reverse the flow of a river. We have covered thousands of acres of Illinois soil with concrete, and we've created our own version of nature, complete with "hills" and "lakes," in our subdivisions and parks. We have created and released pollutants and introduced millions of exotic visitors—cats, dogs, starlings, carp, and Eurasian and Mediterranean plants, among others—into the landscape. Nationally, Illinois now ranks a dismal forty-ninth in the amount of intact natural land.

DIVERSE AGRICULTURAL LANDSCAPE

Yet our landscape transformation has not occurred because of some statewide blueprint or systematic plan. Instead it happened piece by piece—a wetland drained here, a patch of grassland plowed up there, a few more acres paved over every month. As William R. Edwards observed in *Man, Agriculture, and Wildlife Habitat—A Perspective,* "We tend to think of wildlife habitat in a 'then and now' context—then we had it, now we don't. We have been losing habitat for a long time and the effects are cumulative. . . . Because changes in habitat have accrued slowly relative to our individual perspectives, we have been slow to notice their cumulative effects."

As we begin the twenty-first century, we have gained an unprecedented understanding of these cumulative effects on wildlife. The good news is that much of the patchwork elimination of habitat can be reversed. To some degree, the landscape can be restored and reconstructed piece by piece—if Illinois landowners all do their parts to reverse the 200-year-old trend.

Those who plan to join this great protection and restoration effort must understand how our various wildlife populations reached their current status and where they're headed. This chapter provides some insight.

COTTONTAIL RABBIT

## *How Did We Get Where We Are Today?*

Human history is inextricably linked to the state's natural resources. Without the bountiful flora and fauna, the fertile soils, and the extensive network of rivers and streams, our history books would tell different tales. And yet people across the generations have typically taken these resources for granted. For the most part, the decisions humans have made regarding the landscape haven't been intentionally malicious. Pioneers on the land that became Illinois reaped harvests of plants and animals. They believed these resources were truly infinite. Who in 1800 could have imagined that, two centuries later, twelve million people would call this state home?

From the early pioneer settlement period through about 1900, humans made use of wildlife, in numerous ways, from food to fashion. There were no legal limits on the number of ducks, rabbits, deer, and other game animals that could be taken for the dinner plate or the marketplace. Egrets and herons were killed by the thousands in the late 1800s, their feathers sought for adorning hats and purses. Fur-bearing animals, including beavers and foxes, supplied a booming fur trade.

Around the turn of the century, some people became aware of and concerned about the dramatic loss of wildlife. Conservation organizations and agencies were created, and their members and staff worked diligently to regulate the harvesting of wildlife. For some species, such as the passenger pigeon and Carolina parakeet, the protection came too late. But for others, like the white-tailed deer and wild turkey, both close to extirpation, the concerted efforts eventually paid off.

As laws were being written to regulate wildlife harvesting, an even more serious threat emerged. The animals themselves would now be protected from overexploitation, but their homes were not. The numbers of humans and their technical capabilities were rapidly increasing; this meant more and more extensive use of the land and its flora for human purposes. While most wildlife species could potentially have recovered from the problem of over-harvesting, the effects of altered and lost habitat would prove to be much more disastrous.

During the time that wildlife were being taken without limit, plant resources, especially trees, also were being exploited. It is estimated that during the 1800s, Illinois lost two-thirds of its forests. And after the self-scouring steel plow was invented by John Deere in 1837, nearly all of the state's prairies were transformed into crop fields and pastures.

Wetlands, too, began a downward spiral in the 1800s. The first Illinois drainage district was formed in the mid-1870s. By the early 1900s, many prairie wetlands had been drained, and bottomland swamps and floodplain wetlands had been leveed off and drained. Pollution in many forms, such as untreated waste from Chicago that was dumped into the Illinois River, began infiltrating the river's wetlands in the late 1800s.

It wasn't until the late 1800s that concerned citizens began to develop the concept of land preservation. Exploitation of the land was taking its toll, and with wildlife newly protected from overconsumption, conservation agencies and organizations turned their energies in a new direction. Thus began the effort to conserve Illinois wildlife and natural

Gravel terrace prairie

Henslow's sparrow

resources through habitat protection and management—work that continues to this day.

Government agencies began preserving land through acquisition, while many of the private groups focused on public education. Biologists with the Illinois Department of Conservation (since 1995 the Illinois Department of Natural Resources) realized in the 1940s that with 95% of Illinois land in private ownership, successful conservation efforts would have to include these landowners. The department initiated a program to assist landowners with wildlife habitat restoration on their properties. It seemed that with land acquisition and public education, wildlife populations could only improve.

BLUEBELLS IN BOTTOMLAND FOREST

While Illinois wildlife would certainly be in worse shape today without these efforts, unforeseen factors continued to hamper habitat protection and restoration progress. Conversion of wildlife habitat to human uses continued unabated, far outpacing conservation efforts. Although some prairie-dwelling birds were devastated by the loss of the prairie grasslands in the 1800s, many adapted to the non-native pasture and hay grasses that replaced the prairie. But between the 1930s and 1960s, intensified farming transformed most of these agricultural grasslands to row crops. Planted to corn and soybeans, these acres left little habitat for grassland wildlife. Additional farm policy changes in the early 1970s encouraging maximum land usage for row crops led to even more elimination of habitat. This time the victims were the woody fencerows and other odd areas adjacent to crop fields.

As Illinoisans became ever more mobile with the increased use of the automobile, they required more roads; as their numbers and relative wealth increased, they sought more and bigger residential areas and shopping outlets. Woodlands and wetlands were replaced with concrete and steel. And while industry-generated (point source) pollution had been reduced by regulations, non–point source pollutants, such as agricultural and household pesticides, increased by mid-century.

Today, our burgeoning human population continues to threaten the remaining bits and pieces of wildlife habitat in Illinois. The sheer number of people demanding living space, food, fiber, and recreation sites creates competition with wildlife for the land. And it is not just an increase in the amount of land we humans seek but the pattern of

ILLINOIS MUD TURTLE

our dispersal across the state that is greatly affecting what is left of the natural world. Rather than occupying some sections of Illinois and leaving others wild, we have transformed nearly every region to varying degrees.

The construction of homes and accompanying services and amenities in many rural towns and outlying areas shows no sign of ceasing. As farms compete with commercial and residential development for our finite land base, wildlife and their natural environs seem destined to be eliminated. Humans, rather than being an integrated part of the natural world as we once were, are now truly the dominant force on the landscape.

## *Where Are We Today?*

Before European settlement, the Illinois landscape contained at least eighty-five distinct natural community types within fourteen natural divisions. These communities ranged from post oak flatwoods to dolomite prairie to cypress swamp, each containing a unique plant assemblage that provided habitat for a variety of animals, some of them unique to one or two of these specific communities. Most of these community types exist today, but many have been reduced to small "museum" relics of just a few acres. Others are still fairly well represented on the Illinois landscape but are degraded or fragmented into small, scattered islands. Because of the reduction and fragmentation of our wildlife habitat and the many negative influences impacting existing habitat, our natural communities have lost some of their biological diversity.

LITTLE BLUESTEM ON SAND PRAIRIE

These factors—less habitat, poorer quality habitat, habitat fragmentation, fewer habitat types, and reduced biodiversity within habitats—are the primary problems faced by Illinois wildlife today.

## *Quantity*

The acreage of every habitat type except cropland is a small percentage of what it was 200 years ago. These decreases have correspondingly reduced the numbers of many Illinois species. Some species that inhabit specialized or historically less abundant habitats (for example, those that occupy sand prairies, like the state-endangered Illinois mud turtle) have suffered extreme declines or disappeared altogether. Likewise, some historically abundant species that require large tracts of one habitat type have declined as a result of elimination or fragmentation of their habitats. Examples are upland sandpipers, which need expansive grasslands, and ovenbirds and hooded warblers, which need large forest tracts.

Many species have adapted to our human-created landscape, and some are believed to have even larger statewide populations now than existed 150 years ago. The red-winged blackbird, the horned lark, and the white-tailed deer are examples of such adaptable species.

## *Quality*

Many existing wildlife habitat parcels, both large and small, are subject to a variety of negative human influences, thus reducing their benefits to wildlife. An analogy could be drawn to an aging city. There may be plenty of housing units and homes available, but many have deteriorated to the point that they're uninhabitable. The invasion of exotic plant species, increased populations of non-native predators (especially domestic dogs and cats), pollution, mowing, excessive logging, increased noise, the presence of humans—all of these contribute to an inhospitable environment for wildlife. While some species tolerate such negative influences, or are seemingly unaffected by them, most wildlife survive best in habitats more closely resembling those of pre-European settlement, in locations with few or no human intrusions.

Barn owl

## *Fragmentation*

As human dispersal and intensive land use patterns continue, the pieces of remaining quality habitat have come to resemble islands isolated in oceans of human-altered land. While some species are not affected by this fragmentation, a great many are. This is especially true for the less mobile species—amphibians and certain reptiles and small mammals. If a given parcel of habitat becomes too small and isolated, these species cannot perpetuate their genetic diversity and will eventually disappear. And even the more mobile species are increasingly susceptible to premature death, when they are forced to travel between natural habitats without protective cover.

The fragmentation of larger tracts into smaller ones allows predators, both domestic and wild, easier access to species that use

RED-WINGED BLACKBIRD

the habitat interior, thus altering the predator–prey balance. Forest fragmentation creates a higher proportion of forest perimeter to forest interior, allowing the edge-dwelling brown-headed cowbird, a nest parasite, more access to nests of other birds in which to lay its eggs. Most Illinois forest-nesting birds have not evolved with the cowbird and do not effectively deal with its parasitism. And small forests and grasslands do not provide optimum cover from severe weather, such as high winds and deep snows, because these environmental elements can more readily penetrate their interiors.

## *Reduced Biodiversity*

While the human-generated factors affecting habitat may have decreased the populations of many wildlife species, research has shown that for some groups of animals, such as birds, the statewide population has held fairly stable this century. What has changed is the composition of the animal community. Where there used to be a large diversity of birds and other wildlife, now there are large numbers of a few species—those readily adapted to the "new" landscape—and small numbers of many other species tenaciously, and sometimes tenuously, hanging on. A good example of this change can be found in the results of the annual Illinois Spring Bird Counts, where the numbers of both individual birds and bird species are compiled. The past several years have shown that out of the more than 265 species recorded, the ten most common—among them red-winged blackbirds, Canada geese, and the non-native house sparrows, starlings, and pigeons—make up 45% to 50% of the total numbers of birds counted.

The loss of biodiversity in Illinois and around the world is of grave concern to biologists and conservationists. Nature is based on an intricate web of interconnected and interacting organisms. It may have been a surprise to you to learn in the introduction that Illinois has more than 27,000 mammals, birds, reptiles, amphibians, fish, insects, mussels, and other invertebrates within its borders. And more than 3,200 plant species have been recorded in the state. The average Illinois citizen may be familiar with only a few native species. But just because the other creatures or plants aren't seen or familiar to most of us doesn't mean they are not important to the Illinois environment. Unfortunately, many species that fill a specific niche or provide an important food source for other animals disappear with little human notice. It is especially important that we reduce our systematic elimination of insects and other invertebrates to keep the web of life healthy in Illinois and to keep intact populations of what many consider their favorite wildlife—game species, songbirds, and the like. It is also important to protect our remaining natural communities and to restore and reconstruct the complex mosaic that once composed the Illinois landscape.

*The Changing Illinois Environment: Critical Trends,* a report produced by the Illinois Department of Natural Resources and The Nature of Illinois Foundation, summarizes the dilemma succinctly: "There is evolving a trend toward a generic Illinois environment, populated mainly by 'generalist' species able to exploit simplified ecosystems." We need to take action to conserve our state's biodiversity, because the few species that are adaptable, such as white-tailed deer, Canada geese, grackles, and red-winged blackbirds, have come to typify our state's fauna, and vegetation such as bluegrass, fescue, corn, and soybeans have become the dominant plants—a relative landscape monoculture, considering Illinois' rich natural heritage.

## *Where Are We Headed?*

So what's wrong with deer, red-wings, and Canada geese? Actually, these native species are valuable, but they have done so well that in some places they're regarded as nuisances. In healthy ecosystems, nature takes care of surplus animals and plants and keeps the populations of others healthy enough to ensure long-term presence. Our goal should be to reconstruct, restore, and protect healthy ecosystems so that nature can function as close as possible as it did prior to presettlement times.

Today, land-preservation efforts continue, with the goal of protecting high-quality remnants of Illinois' historic natural communities. Governmental agencies and private groups have also saved thousands of acres of other habitat from the saw and plow, the homesite and pollution. State and federal biologists, hunting and fishing groups,

MALE CANADA GOOSE GUARDING NEST

birdwatching and hiking enthusiasts, land-protection groups, and other conservation organizations and advocates continue to effect positive change on the landscape. A recent trend has emerged to address the need for larger blocks of quality habitat. Projects are underway to convert sizeable tracts of marginal farmland to wildlife habitat, such as with the 10,000-acre Emiquon National Wildlife Refuge on the Illinois River near Havana and the Cache River project in southern Illinois. The closing of military properties such as the Joliet Arsenal and the liquidation of utility and mining company lands—and subsequent transfer to state and federal conservation agencies—have provided thousands of acres of land that are now being restored or reconstructed as wildlife habitat.

But the land protected and managed by government and nonprofit groups is merely a teaspoonful in a bathtub of water. Private landowners hold the key to the future of Illinois wildlife. Unfortunately, though, private lands will continue to be a focus of competing land-use issues. The mass conversion of woodlands to homesites near cities and towns threatens every corner of Illinois. Although regulations restrict destruction of wetlands, they continue to be drained and filled for shopping areas and residential developments. And the associated activities that accompany development, such as destroying woodlands to build homes and ponds, providing trails for off-road vehicles, and introducing pets, further stress the landscape.

While we're unlikely to see a significant decline in Illinois' human population and its demands, the future for our state's wildlife need not be hopeless. If residents can be educated and motivated to address the pertinent problems, most wildlife species stand a reasonable chance of continuing to reside in Illinois. The Illinois Department of Natural Resources' Acres for Wildlife program has long been popular; hopefully the number of landowners committed to managing their property for wildlife will continue to grow.

SHADOW OF GARDEN SPIDER ON PRAIRIE DOCK LEAF

The exploitation of natural resources in our state points to no one generation or segment of the population. Activities have occurred as a result of each generation's knowledge and needs. *The Changing Illinois Environment* illustrates this point: "Perceptions about the Illinois environment vary enormously across eras and social groups. Change begets change; as each generation of Illinoisans alters the landscape, later Illinoisans see and feel differently about the land, and thus act differently regarding it."

Because we have begun to appreciate the value of our flora and fauna and we understand the importance of balancing our needs with those of the natural world, the responsibility for caring for wildlife lies firmly in our hands.

## Suggested Reading

*A Sand County Almanac with Essays on Conservation.* A. Leopold. 2001. Oxford University Press, New York, NY.

*A Sand County Almanac and Sketches Here and There.* A. Leopold. 1987. Oxford University Press, New York, NY.

*Comprehensive Plan for the Illinois Nature Preserves System. Part 2: The Natural Divisions of Illinois.* J. E. Schwegman. 1973. Illinois Nature Preserves Commission.

*Endangered and Threatened Species of Illinois: Status and Distribution. Volume 1: Plants. Volume 2: Animals.* Edited by J. R. Herkert. 1991. Illinois Endangered Species Protection Board.

*Habitat Management Guidelines for Amphibians and Reptiles of the Midwest.* B. Kingsbury and J. Gibson. 2002. Partners in Amphibian and Reptile Conservation.

*Man, Agriculture and Wildlife Habitat: A Perspective.* W. R. Edwards. 1985. Illinois Natural History Survey.

*Prairie Establishment and Landscaping.* W. E. McClain. 1997. Division of Natural Heritage Technical Publication #2, Illinois Department of Natural Resources.

*Strategic Plan for the Ecological Resources of Illinois.* G. Bonfert. 1995. Illinois Department of Natural Resources, Illinois Endangered Species Protection Board, and Illinois Nature Preserves Commission.

*The Changing Illinois Environment: Critical Trends. Volume 3: Ecological Resources Summary Report of the Critical Trends Assessment Project.* 1994. Illinois Department of Natural Resources and The Nature of Illinois Foundation.

*The Crisis of Wildlife Habitat in Illinois Today.* Illinois Wildlife Habitat Commission Report, 1984–1985. Illinois Department of Natural Resources.

*The Essential Aldo Leopold: Quotations and Commentaries.* Edited by C. Meine and R. L. Knight. 1999. University of Wisconsin Press, Madison.

*Where The Sky Began: Land of the Tallgrass Prairie.* J. Madson. 1995. Iowa State University Press, Ames.

*Chapter 2*

# *Understanding* the Natural World

*Preparing a wildlife habitat management plan without having a basic understanding of wildlife and plant ecology could be compared to doing your own car repairs without knowing how an engine works—you might luck out, but you're more likely to run into trouble. This chapter explains fundamental ecological concepts to lay the foundation for sound habitat management.*

GREAT BLUE HERON
PREVIOUS PAGE: LEFT, MONARCHS RESTING DURING MIGRATION; RIGHT, NORTHERN BROWN SNAKE

## *How the Natural World Is Organized*

The management activities that biologists recommend to help our natural world are based on ecology—the scientific study of the interactions and interrelationships of plants, animals, and their environments. In keeping with the human need to organize the way we view the world, the scientific community has defined categories for how the natural world is organized and functions. The broadest unit is the ecological system, or ecosystem, a term used in the field of ecology for the last hundred years. An *ecosystem* consists of all the plants, animals, and other living organisms, plus their environment, in a given area. The *abiotic resources* are the nonliving portion of the ecosystem, made up of physical components such as air, water, soil, bedrock, and climate.

The living part of the ecosystem is referred to as the *biotic community,* or the groups of organisms living together in a specific area. The terms *plant community* (or vegetative community) and *animal community* refer to specific groups of organisms. Communities are made up of collective populations of plants and animals. A population is all the individuals of one species.

Wildlife management is conducted at each of these levels: the ecosystem level, the community level, and the population level. For example, the manipulation of water depth in a wetland, by altering both the living and nonliving parts of the wetland, would be considered ecosystem management. Still larger scale ecosystem management would involve an entire watershed. An example of community-level management is prescribed burning on a prairie. Such burning is conducted specifically to manipulate plant communities. White-tailed deer are managed at the population level in Illinois by regulation of their annual harvest. Another recent example of population management is bringing in prairie chickens from other states to increase the diversity of the gene pool in our endangered Illinois flocks.

To be effective, wildlife management must be based on the systems and processes inherent in nature. Understanding ecosystems is key to successful wildlife habitat projects.

## *Understanding Ecosystems*

One of the more confusing aspects of understanding ecosystems is size: How big is an ecosystem? Actually, an ecosystem can be any size. A pond can be an ecosystem. So can a forest. A crop field can be an ecosystem, as can a cliff, a cave, or a backyard. But an ecosystem can also be huge—the Yellowstone ecosystem, for instance, covers parts of three states and encompasses a variety of community types. An example in our state is the Illinois River ecosystem, which includes the river and its entire watershed. Yet there are many

HERON POND IN SOUTHERN ILLINOIS IS A BALDCYPRESS-TUPELO SWAMP ECOSYSTEM.

smaller ecosystems within this larger one. The bottomland forests and the backwater lakes found along the river can be viewed as individual ecosystems.

The concept of ecosystems is broad so that it can be used in a variety of circumstances to study and manage the natural world. For property owners, the ecosystems most important to consider for wildlife are grasslands, woodlands, wetlands, and croplands. Later we will discuss how these relate to the larger landscape of which a given piece of property is a part.

Be aware that there is some misused terminology in the vocabulary of modern wildlife management. The terms *ecosystem*, *vegetative community*, and *habitat type* are often used interchangeably. In this book, discussion of "habitat type" generally refers to a community dominated by a group of plants (woody for woodlands, herbaceous for grasslands) or a community defined by a dominant physical characteristic, such as the presence of water in an aquatic habitat, or by intensive disturbance, as in croplands. Often biologists will also use the term *ecosystem* as they do *habitat*, and they will further be referring to a community type. You might hear a prairie called a prairie ecosystem, a prairie community, or a prairie habitat. Since a prairie can be all of these things in the purest sense of their definitions, all of these usages are correct.

INTERSPERSED CROPLAND, WOODLAND, GRASSLAND, AND AQUATIC HABITATS

There are a number of important characteristics to understand about ecosystems. One is that all their components, both living and nonliving, are connected and dependent—to a greater or lesser extent—on each other. For example, if you remove a large predator species, like a wolf, from an ecosystem, a major shift will result both in the prey populations and in the vegetation and the other creatures that the prey feed on. The chain reaction continues as the vegetation is altered and populations of small invertebrates change, which affects in turn the soil and water.

Ecosystem complexity is also important to understand. An ecosystem can be fairly simple, like an agricultural field that contains only one planted crop, a few weeds, and a handful of animals. In this type of system the interaction between plants, animals, and physical components is relatively simple. But natural ecosystems that have been less altered by humans usually have a much larger variety of plant and animal species. The original Illinois prairie ecosystem, for example, contained as many as 300 plant species. The variety of an ecosystem's biotic community is often referred to as its biodiversity. The more biodiversity within an ecosystem, the more numbers and kinds of plants and animals are found there.

In wildlife habitat protection and creation, maintaining or enhancing ecosystem complexity is usually a primary goal. A complex ecosystem usually attracts and supports more wildlife. It is generally more resilient to damage or complete destruction by uncontrolled fire, disease, storms, and other disturbances, both natural and human-induced. It is important to understand, however, that patience is paramount in habitat restoration work. Habitat establishment aimed at re-creating a complex ecosystem takes time. Establishing vegetation is only the first link in a chain of events that will eventually result in high-quality wildlife habitat. Soil microbes need to establish; insects and invertebrates need to move in. In time, animal species dependent on them will appear.

WHITE-TAILED DEER

All species contribute something toward the healthy functioning of an ecosystem. But too often people have decided arbitrarily to eliminate some part of an ecosystem without regard for its importance to the whole. A classic example has been the systematic destruction of snakes—even by those who profess to be conservationists. Yet snakes play a very important role in every ecosystem where they live. Many landowners try to invite wildlife to their property, but if an undesired creature shows up, it is killed or driven off.

There are instances in which wildlife can threaten human-built structures—woodpeckers and wood-sided homes, for example. If an animal is causing specific damage, control may be necessary, although methods other than killing are usually possible and preferable. It is common when landowners are trying to establish habitat that certain species will hamper their efforts. For example, rabbits and deer may feed on newly planted trees, shrubs, and wildflowers. Beaver will sometimes cut down forest plantings or retard the establishment of woody cover in stream-bank stabilization projects. In these types of cases, some sort of repellent or control may be necessary.

The rule of thumb is this: Never eliminate wildlife just because you don't like it. If an animal causes serious damage or poses a serious health hazard, first consider methods of discouraging it. If this fails and removal or elimination is necessary, consult an Illinois Department of Natural Resources biologist. Most animals are protected by law, and killing an animal without authorization can result in a stiff penalty. Wildlife populations can often be managed through regulated hunting and trapping so they don't become a nuisance.

Remember, each species has an important role in nature, and if you're interested in building and maintaining the best wildlife habitat possible, you need to make room for all wildlife in the ecosystem.

Ecosystems are subject to a variety of processes. Rather than being static, ecosystems are dynamic. Change is always occurring as plants and animals struggle among themselves and with other plants and animals for survival. Factors such as weather, fire, and floods also create change in ecosystems. It is important to understand, too, that not only do changes in the physical environment affect the biotic community, but the reverse is true as well—changes in the plants and animals can affect their environment. A significant fluctuation in the plants or animals inhabiting a pond can change the water's chemistry—a massive die-off of algae in a pond can deplete oxygen, for example. Or if a disturbance removes a large portion of tree and ground cover in a forest, sunlight penetrates to the newly exposed soil and raises the ground-level temperatures. In both of these instances, the change in the physical environment would spur further change by affecting the biotic community—the pond inhabitants in the first example and the soil and the leaf-litter dwellers in the second.

Change of any kind in the biotic community or its environment creates changes in the whole ecosystem. Change can be favorable or

Edge of a natural lake ecosystem

TWELVE-SPOT DRAGONFLY

unfavorable depending on the species in question. Usually an ecosystem fluctuation favors some plants and animals and not others. However, natural change in ecosystems should not be viewed overall as a negative process.

Humans have caused great changes in Illinois ecosystems, often to the detriment of individual ecosystems. But human-created disturbances can also be used to undo some of the damage we've done. Our ability to effect change in portions of an ecosystem for the benefit of wildlife is at the heart of this book. A landowner can establish specific plants on a site, manipulate water levels in a wetland, or introduce fire—all of which, if done for a specific purpose, can improve conditions for wildlife.

Let's address more specifics on the biotic components in ecosystems.

## *Understanding Wildlife Populations and Communities*

To begin understanding the life requirements of wildlife, we need look no farther than the needs of a familiar animal: the human. Just like us, wildlife require four basic elements to live: water, food, shelter, and space. Food and water are prerequisites to sustaining life; shelter—from weather, from enemies, and for raising young—perpetuates life. And all animals need physical space for themselves and their families, although the amount needed varies among species.

Habitat is what provides all of these elements for wildlife. The term habitat is used in several ways, but it is basically defined as a living environment. The term can be used singly (an animal's habitat) or collectively (wildlife habitat). The word is also used to define a type of living environment. Our management recommendations address four predominant Illinois habitat types: grasslands, woodlands, wetlands, and croplands.

Some wildlife species may be able to satisfy all four basic needs in one type of habitat. Other species have complex needs and require several habitats to secure adequate food, water, shelter, and space. The collective area needed by an individual animal to meet its normal living requirements is known as its home range.

An animal's home range should not be confused with its natural range, which refers to the geographic area in which the species can be found. For example, the natural range of the eastern chipmunk includes eastern Canada and the eastern half of the U.S. But an individual chipmunk's home range is from less than half an acre to three acres, depending on local conditions.

MUSKRAT

RING-NECKED PHEASANT

Home-range size applies to individuals within a species and varies between species. For some, such as the white-tailed deer and the great-horned owl, a home range may encompass one to four square miles. The home ranges of other animals, such as the short-tailed shrew and the spotted salamander, may be just a quarter of an acre. Still other species, including migratory hummingbirds and warblers, have home ranges that may vary seasonally and may be located on two continents. The home range may also vary slightly for any species depending on the quality of available habitat.

Here's an illustration of a species' home range: The ring-necked pheasant has needs that differ between seasons, and to meet those needs it must use several habitats. During much of the year, this bird can be seen roaming the open agricultural fields of central and northern Illinois. For breeding and nesting, however, the female needs to find an undisturbed (unmowed) grassy field that will hide her eggs and provide insects for her newly hatched chicks. And during severe winter weather, the birds need thick, erect vegetation, such as cattails, prairie grass, or brushy cover, for shelter. The area required to meet these needs, generally one square mile in Illinois, is the pheasant's home range. However, individuals have been known to disperse as far as 15 miles when needed to fulfill their living requirements.

When an element necessary to a species' survival is in short supply or completely absent, it is referred to as a limiting factor. Because wildlife have complex and varied needs, the lack of any particular element can limit population size or even prevent a species from inhabiting an area. For instance, a woodlot may be filled with mast-bearing oak and hickory trees that produce enough food for twenty squirrels. But if the woodlot contains only a couple of trees with cavities, the squirrel population will be limited by the availability of secure sites for raising young. The area probably cannot support twenty adult squirrels because of a limiting factor—insufficient nesting sites.

Most of today's limiting factors for Illinois wildlife are caused by habitat degradation and elimination. Natural occurrences such as weather extremes, disease, and lack of food can also be limiting factors.

Some landowners become disappointed that habitat development or enhancement doesn't produce immediate results, or the results they expected. They wonder why their grass planting has failed to attract quail or why they don't have wood ducks in their newly created marsh. The reason is probably a missing habitat element or an unapparent limiting factor that has prevented the local population from expanding into the landowner's "new" habitat. Landowners should not let such a situation discourage their habitat development; improvements targeted at a specific species often work with time, provided the property is within the animals' normal geographic

NEWLY HATCHED RING-NECKED PHEASANT CHICKS

range. Sometimes a particular limiting factor will be eliminated after habitat plantings have become well established. For example, the simple presence of grass may not be enough for certain grassland birds. Some species, such as the Henslow's sparrow and the sedge wren, need undisturbed grassy areas that have accumulated a couple years of dead grass thatch. A first-year planting cannot provide this.

Ecosystems are dynamic, and the populations within them fluctuate from year to year. But on average, all habitats can support only so many animals of any given type. This limit is defined as an area's carrying capacity. When too many animals of a species are born into or immigrate into an area that is at its carrying capacity, some animals will have to go, either by leaving or by dying. The carrying capacity for any species varies seasonally and yearly, depending on the presence or absence of limiting factors. During a drought year, for example, fruit-bearing trees and shrubs may produce less fruit than usual, and the carrying capacity of the drought area would likely be reduced for the wildlife species that subsist on the fruit.

Many species produce an overabundance of offspring each year to ensure an adequate survival rate. If "new" suitable habitat has been created near existing habitat, some of the surplus animals will move into the new area. But if there is no new habitat for dispersal, the surplus animals are eliminated through starvation, predation, or disease. Hunting and trapping are regulated to use the surplus without destroying the base population.

The principles of carrying capacity continually determine population sizes and viability. Landowners trying to increase wildlife populations must understand that wildlife cannot be stockpiled; populations will always be limited by the amount and quality of habitat available to them.

A RAILROAD PRAIRIE PROVIDES PHEASANT HABITAT.

## *Understanding Plant Populations and Communities*

Although the biotic community includes all the living organisms, plant and animal, occupying a particular area, it is often referred to by its dominant vegetation. For instance, in Illinois you might hear of an oak–hickory forest community, a cattail marsh community, or a prairie grass community.

However, while communities are generally identifiable, the landscape doesn't always consist of neatly defined parcels. Where different vegetative communities meet there is often some overlap, creating a transition zone called an ecotone. Ecotones are often subject to great fluctuation because disturbances usually affect the two adjoining communities differently. For example, the ecotone between a forest and a grassland would contain some shrubby plants and maybe a few scattered trees, with lots of grassland plants beneath them. If a fire were to burn the grassland and the edge of the woods, some of the woody plants in the ecotone would be destroyed and the grass would flourish. The edge of the woods would be opened to sunlight, creating more favorable conditions for the grassland plants. The change would allow the grassland to advance a little into the forest.

While ongoing change is not always apparent, plant communities are dynamic rather than static entities. All land in Illinois, whether backyard, corn field, or bottomland forest, is subject to a process called succession. Succession describes the continuous, fairly predictable change of plant species and plant communities over time. The ongoing competition among various plants and plant communities to establish themselves on the landscape is why succession occurs. How it occurs is a matter of biology. Different types

GRASSLAND–WOODLAND ECOTONE

of plants have different life cycles and adaptations for surviving in the environment, and over time some plant species and groups of plants will naturally be better adapted to local conditions.

When does succession occur? It happens continually on a small scale. Large-scale succession of whole plant communities, though, occurs whenever there is a major change in an area's environment. Natural disturbances such as floods, tornados, and severe icestorms can damage or destroy an existing community, and a major successional course usually begins.

In Illinois, the human activities most often leading to succession of plant communities are abandonment of agricultural fields and logging. Yet the process of succession can be observed right outside our homes. Any time a tree pops up in an unmowed area or weeds appear in a newly plowed garden, succession is in progress. What keeps our backyards from becoming woodlands are lawnmowers, herbicides, and garden tillers.

If it is allowed to proceed without disturbance, plant succession has fairly predictable stages and outcomes. Generally, in much of Illinois today, the process of succession in a bare field starts with the pioneer stage of annual plants, which eventually are largely replaced by perennial herbaceous plants, then shrubs and trees, and ultimately a mature forest. Here's an example of what could predictably happen if an upland central Illinois crop field were permanently abandoned, with no further human interference: First it would be invaded by herbaceous annuals such as foxtail and ragweed. After two or three years, perennial herbaceous plants like goldenrod and milkweed and grasses like smooth brome or bluegrass would get firmly established, outcompeting and replacing many of the annuals. Assuming a local seed source was present, after another two to three years, woody species such as dogwood and elderberry shrubs and seedlings of invader trees such as silver maple, eastern redcedar, mulberry, and box elder would begin appearing. As the woody species developed extensive root systems and increasingly shaded the ground, they would dominate the site. As the invader trees grew, a young forest would develop. Eventually (in twenty to forty years), the trees would sort themselves out, and certain species would dominate to form the canopy of a maturing forest.

**Figure 2.1 Succession of Representative Volunteer Plant Species on Well-Drained Soil**

Edge habitat between woodland and cropland

Most of Illinois' openland habitats—grassland and cropland—would ultimately revert to woody cover if left undisturbed. Which plants would appear during the course of succession would be dictated first by what seed sources were in the area and second by the specific physical conditions of the site (drainage, soil type, slope, etc.).

When disturbance is introduced during succession, the process can be either retarded or encouraged. Much of the Illinois land used for agriculture is intentionally kept at an early successional stage by periodic tillage. For the landowner trying to reconstruct some type of habitat, such as a woodland, minimizing tillage, mowing, and other disturbances should encourage successional progress to that targeted community type. For other habitat reconstructions, such as grasslands, periodic disturbance (fire, mowing, discing, herbicides) is necessary to prevent woody encroachment and to help the herbaceous plants flourish. Strange as it may sound, many mature communities have evolved to actually depend on certain disturbances to prevent succession. For example, many oak–hickory forests need fire to keep them from succeeding to a forest of maple and other shade-tolerant species. Many grasslands need fire to prevent them from succeeding to woody cover. Bottomland wetlands need periodic flooding to flush accumulated organic debris so they do not succeed to woodlands.

Manipulating the succession of habitat, by either retarding it or enhancing it, is the basis of much of wildlife habitat management. But it is important to remember that even if you're managing for an eventually "stable" community, there will always be some ongoing successional change. You can't freeze nature into an oil painting of exactly the plant species you want, just where you want them. This doesn't mean that you won't eventually have a prairie or woods that seems relatively unchanged over the years. What it does mean is that certain plants you introduce during establishment of your habitat may, for one reason or another, not be able to persist or compete with the others and will eventually disappear.

Here are some additional points to be aware of about succession:

- It is easy to create early successional habitats simply by applying some sort of disturbance. But we cannot reconstruct a mature, later-successional-stage community in a short period. For example, we must simply wait out the time necessary for trees to grow to their mature state. We can, though, have limited impact on the efficiency with which that growth happens. For example, we can plant desirable trees such as oaks and hickories and at the same time suppress shade-tolerant trees to give the slower-growing oaks and hickories maximum growth opportunities. Given the lengthy time it takes to re-create mature communities, especially the forested types, it should always be a priority to preserve those that already exist.
- The process of succession in Illinois has been affected by the widespread establishment of non-native plants, which sometimes outcompete native species. As a result, we must often "assist" in succession if the goal is to establish a diverse, specific community of native plants. Fescue, for example, can become so thick that it prohibits the establishment of most woody vegetation. And aggressive exotics such as Tartarian or Amur honeysuckle can suppress native, slower-growing trees such as the oaks. These exotic shrubs can quickly dominate and shade the forest floor and prevent sun-loving oak seedlings from establishing.
- Since wildlife have specific habitat needs, succession of plant communities also results in some continuing change of corresponding wildlife communities.

Because controlling or encouraging succession is pivotal in wildlife habitat management, chapters 4, 5, and 6 discuss succession for each habitat in detail.

SAVANNA-RELICT BUR OAK

## *The Bigger Picture: Ecosystems on a Landscape Level*

FRAGMENTED WOODLAND

Understanding how plant and animal populations function and interact with their environment builds a foundation for good habitat management. The next step is to translate that knowledge from these pages into the real world—the everyday management of your parcel of land. But for wildlife the "real world" isn't defined by legal property boundaries. In our state, their reality is the collective landscape we call Illinois, an intricate mosaic of overlapping home ranges of various animal species. Since wildlife operate within their own boundaries and not those of humans, we must look at the bigger picture to make the best land management decisions.

We have discussed habitats as individual units of management. When we look at the landscape, these units form various patterns. While the quality within these units is important, the size and arrangement of these units on the landscape also affect their suitability for various wildlife species.

*Interspersion* refers to the proximity of different habitat types. The interspersion of habitats is directly related to the size and arrangement of the units, and it affects what animal species will occupy them. Several small tracts of different habitat types in close proximity means there is high interspersion, and vice versa.

Interspersion is an important consideration in habitat management planning. Some species may require large, unfragmented tracts of a single habitat type (no interspersion). Other wildlife may need several habitat types (high interspersion), all of which can be small patches but that must be spaced close together. Many species thrive in edge habitat, which is created with higher interspersion. Edge habitat—which exists where two habitat areas meet, such as a brushy fencerow and a crop field or a grassland and a forest—contains animal species from both adjacent habitats and also supports wildlife that prefer edge, such as cardinals, cottontails, and common yellowthroats.

Two illustrations reveal the importance of considering interspersion. Ovenbirds require several hundred acres of unbroken, mature forest for successful breeding, and they have evolved to procure all the necessary elements from that one habitat in the summer. Smaller patches of habitat make them vulnerable to competition from other species, such as the brown-headed cowbird, and to increased predation by abundant edge-dwelling predators like raccoons.

On the other hand, cottontail rabbits need several types of habitat, and because they are not extremely mobile (as most birds are) they need to have the habitats close together. They are often found in edge habitats as opposed to deep forest interior.

Interspersion is not the only consideration that influences habitat suitability. Whether valuable habitats are isolated or connected also determines the landscape's usefulness to most wildlife species. For example, let's say a thirty-acre woodland sits in the middle of 900 acres of crop field, and the nearest woody cover is three miles away. Less mobile species of wildlife, such as reptiles, amphibians, and small mammals, would be unable to travel between the two woodlands. The animals in the thirty acres would be isolated and more susceptible to the effects of inbreeding and adverse weather events. As species disappeared from the island woods, the biodiversity of this forest would decrease and it would eventually support only mobile generalist species, like white-tailed deer and robins.

If, on the other hand, strips or fencerows of woody vegetation existed between the two woodland tracts, genetic intermixing could occur via these corridors. Corridors used to be more numerous in Illinois. The elimination of hedgerows coupled with our continued penchant for "tidy" landscapes has led to the removal of many of our weedy, brushy, and tree-filled windbreaks and fencerows.

Earlier we discussed biodiversity within ecosystems. When we look at the Illinois landscape, biodiversity between ecosystems is also critical. The Illinois Natural Areas Inventory that was conducted in the late 1970s categorized, for the first time, nearly all of the natural community types historically known for our state. Approximately eighty-five communities were recognized. Preserving and restoring this diversity of distinct communities on the Illinois landscape is essential to preserving our variety of wildlife.

## *Putting the Pieces Together*

As you can probably now see, managing for wildlife can include many elements. So how do you, the individual landowner, determine the appropriate habitat to construct or preserve—in the right place and of the right size—for the myriad of Illinois wildlife?

While species are sometimes managed individually, the most efficient way to manage is to consider groups of species that occupy common habitats. This is why most land management revolves around habitat types. Many species use several habitats but prefer one primary type. Generally, if a woodland is managed for woodland wildlife, most species that use this habitat will benefit.

Trying to restore Illinois habitats as we believe they existed prior to European settlement is a way to benefit native species with specific requirements. (See *The Natural Divisions of Illinois* referenced on page 45 for a better understanding of the historic Illinois landscape.) While this book discusses the use

OVENBIRD

of certain non-native plants, the primary emphasis is native plants and plant communities. Where possible and practical, we should use nature as the gold standard to measure our actions.

One of a landowner's most important management considerations is to take a three-dimensional view of wildlife habitat. From the treetops to deep in the soil to the horizon beyond a property boundary—this is the habitat where wildlife live. Habitat types and home ranges transcend our political and legal boundaries and stretch beyond immediate human vision. Try to view the landscape from a wildlife perspective.

SHORT-EARED OWL

Often we act by considering only short-term economic factors. Taking a more holistic view is usually the best approach. Unfortunately, not thinking broadly enough has resulted in a fragmented effort to conserve and restore Illinois' wildlife habitat. But government agencies and private conservation organizations have recently begun to consider a larger "landscape level," or ecosystem-based, approach to their wildlife management activities. Landowners ought to do the same if they desire to restore and retain a diversity of wildlife.

The final point to be learned from lessons of ecology is that wildlife are affected, either negatively or positively, by whatever a landowner does to his or her property—including doing nothing. No Illinois landowner can be passive and claim not to be part of the problem or the solution. The fate of Illinois wildlife rests in the hands of private landowners. With a foundation of basic environmental knowledge, each landowner can make land management choices that build the sustainability of our natural resources.

# *Myth or Fact?*

There are many widely held beliefs about wildlife that are accepted as fact. On the surface these beliefs may seem very logical. However, once you understand how the natural world functions, you question them. Understanding concepts such as plant community succession, basic needs of wildlife, and carrying capacity helps dispel many of these myths. Some common myths and the related facts are detailed here.

## *Myth: Wildlife are starving.*

*Fact:* Wildlife that feed on seeds or grains are far from starving in Illinois. With nearly 80% of our land used for agriculture, and with conservation tillage increasing in our state, food for many wildlife species is abundant in crop fields. And most seed-eaters opt for a variety of foods, including weed seeds and other plant materials. Wildlife can experience hardship in winter when ice or deep snows blanket the ground for days or weeks. But it is the lack of cover, not of food, that usually kills wildlife in the winter.

A different aspect of wildlife food supply is of serious concern. Widespread pesticide use on our yards and farms and loss of vegetation, such as small grains and legumes, have greatly reduced insect populations. Hundreds of bird, mammal, reptile, and amphibian species depend on these invertebrates for food. They are an essential food for young pheasant and Bobwhite quail chicks, and many songbirds eat insects almost exclusively. While many people would argue that we still have more than enough insects, certain wildlife species are likely experiencing a reduction of necessary foods in some instances.

House wren with insect

## *Myth: Hunting and trapping are the reason many wildlife species have declined.*

*Fact:* Hunting and trapping have not contributed to the decline of wildlife since the days of unregulated mass harvesting for markets. Today both activities are carefully controlled by state and federal conservation agencies, based on the principle of carrying capacity. Harvest limits are set in accordance with statistical analyses of species' year-to-year populations. Species that are hunted and trapped are those whose populations are flourishing enough to produce a surplus.

## *Myth: Releasing hand-reared birds builds up local populations.*

*Fact:* Numerous studies have shown that pen-raised birds such as quail and pheasants don't survive long in the wild once they've been released. They don't fear predators, don't know how to seek cover from inclement weather, and may be inept at finding food. And if an area is already supporting all the animals it can (it is at its carrying capacity), adding new individuals only results in a surplus that is eliminated in a few months. Releasing pheasants or quail to supplement a hunting program can be acceptable, but doing it to increase an existing population is a waste of time and money. A better use of those resources would be expanding existing habitat to allow natural population growth.

Some people wonder, if we can successfully raise and release certain endangered species to reestablish a population, why won't it work for all wildlife? Most captive-breeding programs for endangered species work to reestablish an animal to a former range where habitat conditions have taken a turn for the better or where a harmful factor or element, such as the insecticide DDT, no longer threatens a species. Depending on the species, the animals are bred and released into the wild at a young age, or they are raised in a semiwild environment, carefully controlled to simulate a natural situation.

WILD TURKEY

*Myth: Because the state has successfully established new populations of some animals through trap-and-transplant, this should work for any species.*

*Fact:* The wild turkey and the river otter are two species that practically disappeared due to habitat elimination, habitat degradation, and uncontrolled harvesting. Even when these factors improved, physical barriers prevented both species from reestablishing in their former ranges.

The wild turkey was overharvested at the same time its habitat was disappearing. Once harvesting was banned and then subsequently regulated, and some habitat restoration had occurred, the species still wasn't recovering. The birds wouldn't cross towns or large expanses of open agricultural land to repopulate formerly inhabited areas. Biologists assisted by transplanting the wildlife.

The river otter declined initially due to unregulated trapping, and it suffered further declines from the poor water quality that developed in Illinois around 1900. Water quality has improved significantly in some Illinois rivers and streams, but the otters cannot always travel long distances to reinhabit their recently improved former habitats. Transplanting worked because the factors negatively affecting the species were resolved.

There are additional animals that could be successfully transplanted, but only after long and careful evaluation by qualified biologists. Numerous habitat conditions must be met before any species can be reestablished. One or more habitat elements are missing for many of the species that have declined in or disappeared from Illinois, in which case transplanting would not be successful.

## *Myth: The best thing I can do for my land is to let it go "back to nature."*

*Fact:* Nearly all habitats in Illinois today, natural or created by humans, need some form of management.

We have altered the landscape and the dynamics of nature so drastically in our state that most systems are hindered in returning to a native and natural state without our help. Even the high-quality remnant communities scattered around Illinois usually need some assistance to respond to the onslaught of changes we've thrust onto the landscape. For example, we eliminated the natural process of fire, but we now understand its critical role in sustaining many natural communities. And the unabated march of invasive exotic plants into our natural communities will not cease without our help.

TIGER SWALLOWTAIL AND GREAT SPANGLED FRITILLARY ON PURPLE MILKWEED

## *Myth: New and improved varieties of shrubs and trees are better to plant because they grow much faster and provide better habitat.*

*Fact:* The Illinois Department of Natural Resources, along with other agencies such as the USDA Natural Resources Conservation Service, promoted this concept in decades past. However, long-term research has shown that introducing non-native vegetation into native plant communities can have many dire consequences. This has been demonstrated by Tartarian and Amur honeysuckle and multiflora rose. Illinois biologists have not recommended non-native woody plantings for more than twenty years, and they suggest only a few non-native herbaceous plants, such as pasture grasses and legumes, for wildlife. But plenty of non-native varieties of herbaceous plants, shrubs, vines, and trees are still commercially available. Landowners should use only native plants in rural habitat projects, and where possible in backyard habitats as well. Many of our worst invasive exotic species in Illinois—purple loosestrife and garlic mustard, for example—arrived here as flower or herb garden plantings and escaped "to the country" to set up permanent camp. Table 2.1 on page 44 shows some of the problem species that should not be planted, even in backyards.

Even with apparently native plants, careful shopping is required. Plant scientists have made many advances in selecting and developing varieties. The results have brought us "super" varieties of many native species. Switchgrass, a native Illinois prairie grass, is a good example. This grass has proven to be good forage for cattle, and plant scientists have developed more aggressive varieties to improve production. When an improved variety of switchgrass is planted with other native prairie plants, it usually takes over and crowds out the companion species during the first few years.

Problems can arise when landowners buy species that are native to Illinois but whose seed is collected elsewhere. Many landowners shop by mail to save money on prairie plants and seed. It is always better to purchase plants and plant seed derived from stock grown in Illinois. This practice keeps local genetic diversity intact.

## *Myth: There are too many predators.*

*Fact:* When people think of predators, animals such as foxes, coyotes, bobcats, and hawks usually come to mind. Yet many other species are also "predators." Herons are predators of fish, salamanders eat worms, shrews sometimes feed on other rodents, and bluebirds and robins consume beetles, spiders, and other invertebrates. Whatever the species, any wild predator will only increase in proportion to the amount of available prey.

There are two Illinois predators, though, whose abundance doesn't vary with fluctuating prey populations. These animals have increased dramatically in the last several decades, and their disproportionately high numbers on the Illinois landscape are seriously threatening the natural stability of many ecosystems. These "new" non-native predators are domestic dogs and cats. As our human population grows and expands into rural areas, pets have become a dominant force on the land. Unfortunately, their overabundance is impacting some birds, mammals, reptiles, and amphibians. Cats in particular, will hunt whether or not they're hungry. So keeping a pet well fed cannot solve the problem. When possible, landowners should confine pets to a yard area, and put bells on their collars to alert wildlife to approaching danger. Other measures include having pets neutered and controlling the dumping of unwanted animals.

**Table 2.1 Problem Plants***

- Amur honeysuckle
- Asiatic daylily
- Autumn olive
- Burning bush
- Canada thistle
- Chinese pea-tree
- Chinese yam
- Climbing euonymus
- Crown vetch
- Dame's rocket
- European buckthorn
- European water milfoil
- Garlic mustard
- Glossy buckthorn
- Japanese honeysuckle
- Jetbead
- Kudzu
- Multiflora rose
- Musk thistle
- Narrow-leaved cattail
- Phragmites
- Purple loosestrife
- Purple vetch
- Purple wintercreeper
- Reed canary grass
- Round-leaved bittersweet
- Russian olive
- Silver poplar
- Sweet clover
- Tall fescue
- Tartarian honeysuckle
- Teasel
- Tree-of-heaven
- White mulberry
- Winged wahoo

*While some of these species have been commonly used for backyard landscaping and wildlife habitat, they should not be planted.

## *Suggested Reading*

*Comprehensive Plan for the Illinois Nature Preserves System. Part 2: The Natural Divisions of Illinois.* J. E. Schwegman. 1973. Illinois Nature Preserves Commission.

*Illinois Habitat Posters.* [Full-color posters that depict Illinois habitats and the organisms that inhabit them.] Illinois Natural History Survey.

*Illinois Wilds.* M. R. Jeffords, S. L. Post, and K. R. Robertson. 1995. Phoenix Publishing, Urbana, IL.

*Illinois Wildlife and Nature Viewing Guide.* M. K. J. Murphy and J. W. Mellon. 1997. Illinois Department of Natural Resources.

*Technical Reports on Bioregional Resources of Illinois.* [Series covers 21 bioregions of the state.] Illinois Department of Natural Resources.

*The Audubon Society Nature Guides.* [A series of guides published by Knopf and divided by ecosystem: grasslands, wetlands, eastern forests.]

A GARDEN SPIDER PREYS ON A MONARCH.

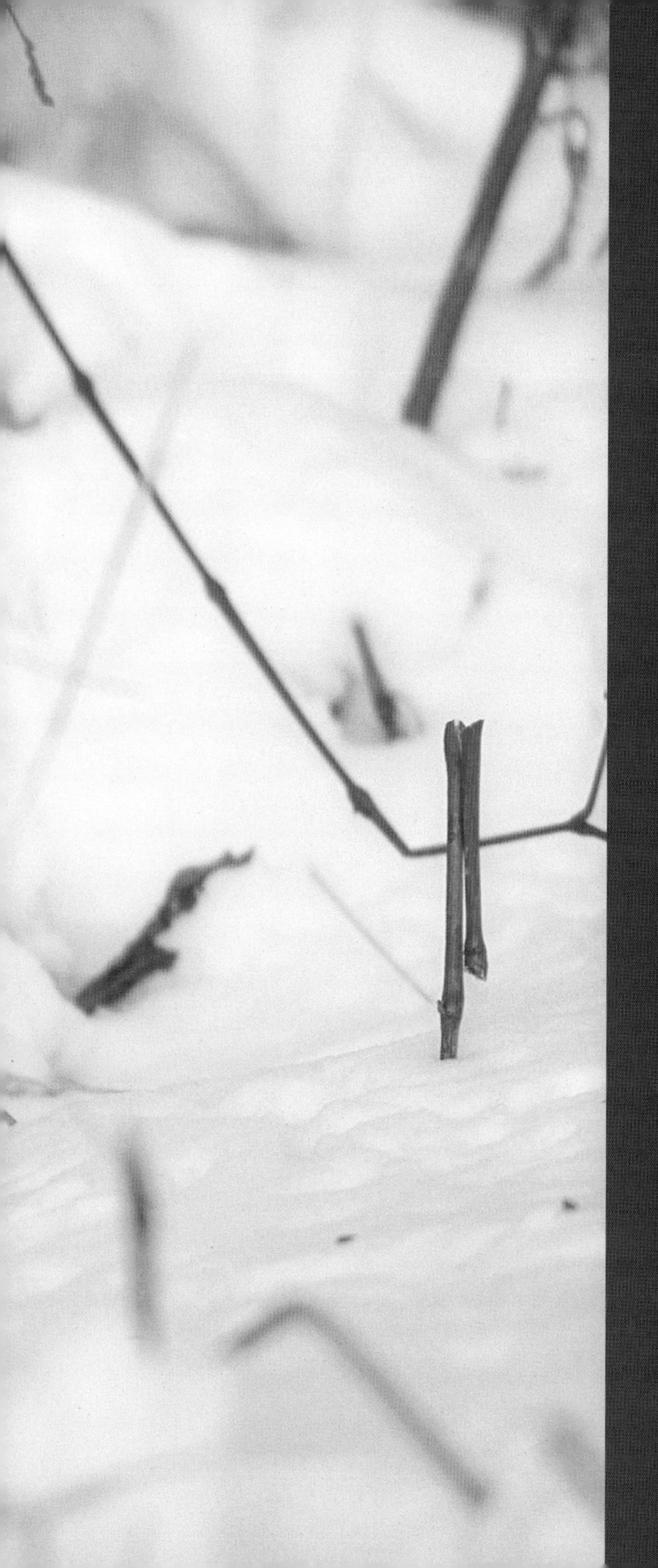

*Chapter 3*

# *Grasslands*

## Including Prairies

*Despite our continual advances, humans seem to have a fondness for old things and olden days. Fortunately for our state's environment, some Illinois residents have developed an interest in a particular feature of days gone by: the prairie. Whether they are planting a garden of native prairie flowers or establishing a patchwork of tall prairie plants on the back forty, Illinoisans are rediscovering the grasslands that inspired our nickname of the Prairie State.*

COYOTE
PREVIOUS PAGE: LEFT, BOBWHITE QUAIL; RIGHT, BISON

Prairie plantings are extremely valuable to wildlife. Of course other grasslands, if managed properly, provide suitable habitat too. Grasslands provide habitat for a variety of wildlife; Table 3.1 (p. 50) outlines the needs of a selected range of animal species and how grasslands meet those needs. In this chapter you'll learn the various types of grassy habitats and how to create and manage them.

## What Defines Grassland Habitat

The defining factor for grassland habitat is, as you would guess, the prevalence of grasses. However, our use of the term *grassy cover* includes an herbaceous broadleaf component. In native prairie these broadleafs are called forbs, and in non-native or introduced grasslands they are usually referred to as weeds or legumes. Fields of grassy cover with scattered trees or shrubs (less than 10% canopy cover by these woody species) are also considered grasslands. Grasslands with 10% or more scattered trees and shrubs are termed *savannas*; they are discussed in chapter 4.

Scientists classify grasslands in many ways. In the native prairie category, for example, there are many *types*, such as hill prairie, wet prairie, and sand prairie, each name referring to a specific defining physical feature of that community. See Table 3.2 (p. 53) for a summary of each type of native prairie and their special management considerations. Fields of non-native or introduced grasses are often described by the predominant grass type, such as a bluegrass pasture or a brome hay field.

In this book we've classified grasslands into three broad categories, based on growing characteristics and the composition of their herbaceous vegetation. *Warm-season grasslands* are dominated by grasses that do most of their growing in June, July, and August. This includes most of the native grasses and many of the forbs of the original Illinois prairies. Typical grass species include big bluestem, little bluestem, and Indian grass. Forbs are numerous; common examples include rattlesnake master, compass plant, blazing star, and spiderwort. There are also native cool-season grasses and forbs in the prairie, but because of our climate they do not form a dominant community type on the Illinois landscape.

*Cool-season grasslands* are dominated by grasses and forbs that grow mostly in the cooler months of spring and fall. Generally the plants in these grasslands are non-native, having been introduced from Europe or other cool climates. They include species common in pastures, hay fields, and lawns such as bluegrass, tall fescue, red-top, smooth brome, and orchardgrass as well as broadleaf plants like alfalfa and red, white, and sweet clovers.

*Old-field grasslands* may contain either cool- or warm-season grasses, but they are usually characterized by a predominance of "weedy" broadleaf plants and develop after abandonment of a formerly disturbed area, such as a crop field or pasture. Typical plants include foxtail, goldenrod, broom sedge, and ragweed. Old fields also often contain scattered woody vegetation, such as dogwood or plum shrubs and blackberry briars.

ROUGH BLAZING STAR IN NATIVE PRAIRIE REMNANT

**Table 3.1 Attributes of Selected Grassland Wildlife**

| Species | Preferred foods | Nesting and breeding habitat needs | Winter habitat needs | Additional notes |
|---|---|---|---|---|
| Rabbit | Spring, summer, fall: wheat, alfalfa, red clover<br>Winter: twigs and bark | Undisturbed grassy cover | Brushy cover, tall grasses and forbs, woodland edges | Brushpiles, brushy fencerows, and clumps are important for escape cover. |
| Bobwhite quail | Young: insects only<br>Adults: legume and forb fruits; grains; seeds; insects, especially beetles | Undisturbed thin grassy cover with good mix of forbs | Brushy cover, tall grasses and forbs, woodland edges | Requires close interspersion of grassland, woodland, and cropland. |
| American kestrel | Small mammals, reptiles, amphibians, birds, insects | Nests in tree cavity or nestbox. Prefers fallow fields and grasslands for foraging. | Prefers crop fields, pastures, hay fields with perches for foraging in winter. | Will readily use nestboxes along roadways. |
| Field sparrow | Seeds of grasses, sedges; insects in season | Undisturbed grassy cover with interspersed weedy or brushy cover | Often moves south of Illinois for winter. Otherwise, prefers tall grass and brushy winter cover. | Needs some taller woody vegetation for song perches. |
| Eastern bluebird | Spring, summer, fall: insects<br>Winter: fruits | Nests in fencepost, tree cavity, or nestbox. Needs areas of short cover for summer foraging. | Usually leaves Illinois for winter. | Steps must be taken to minimize raccoon and cat predation of nestboxes. |
| Prairie kingsnake | Primarily small mammals; secondarily amphibians, reptiles, birds, insects | Prefers old fields | Burrows into the ground for winter hibernation. Will use burrows of mice, crayfish, etc. | Most common in grassy areas along woody edges. |
| Ring-necked pheasant | Young: insects<br>Adults: some insects, grains (corn, milo) | Undisturbed grassy cover | Thick cover such as prairie grass, cattails, plum thickets, pine groves, brushy ravines | While this bird is thought of as a cropland bird, winter and nesting cover are essential to its survival. |

## *Grassland Habitat Issues in Illinois*

When the pioneers began rolling into the land now called Illinois, they were met by a seemingly endless expanse of grass. Indeed, prairie covered about two-thirds of our state. How ironic that today prairie habitat occupies the smallest amount of the Prairie State's acreage. It survives only in abandoned cemeteries, railroad rights-of-way, and a few scattered preserves.

The majority of Illinois prairie was eliminated in a short sixty years, after the invention of the steel plow in 1837. It was a swift, hard blow to some wildlife dependent on this habitat. Fortunately part of the prairie was converted to other types of grassy cover, such as pastures and hay fields. For many wildlife species, like the eastern meadowlark, dickcissel, thirteen-lined ground squirrel, and others, the exchange of one type of grassland for another was acceptable.

Wildlife that depend on grassland habitat face four problems today:

1. The overall amount of grasslands has declined.
2. The size of our existing grasslands and their interspersion on the landscape have changed.
3. Many existing grasslands are of poor quality in terms of plant species and vegetative cover.
4. Grasslands are often managed to the detriment of wildlife. For example, they are mowed or sprayed with pesticides during the critical reproductive period for wildlife during spring and summer.

## *Less Grassland*

Wildlife species that made themselves at home in the "new" Illinois grasslands after destruction of the prairie have been dealt another blow in recent times. Since the 1960s, the acreage of pasture and haylands in Illinois has been steadily decreasing, declining an estimated 50%. Overall, the total acreage of grassland, native and introduced, is only a small fraction of our original twenty-one million acres. There is simply less grassy cover for wildlife to inhabit. The populations of many Illinois grassland birds, such as grasshopper sparrows, eastern meadowlarks, and bobolinks, have declined 75% to 95% in the last half-century.

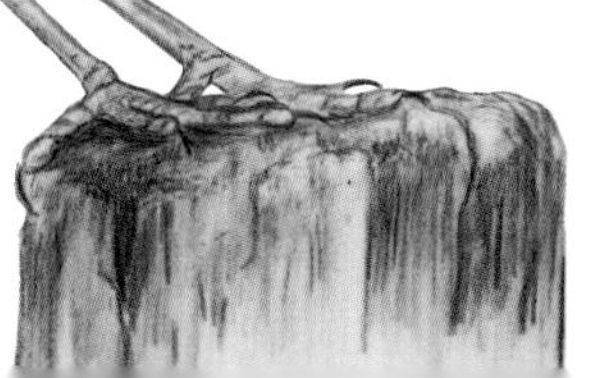

EASTERN MEADOWLARK

TALLGRASS PRAIRIE

TICKSEED SUNFLOWER IN OLD-FIELD GRASSLAND

## *Different Size and Interspersion*

The original Illinois prairie stretched for miles with hardly a tree in sight. Many species, including barn owls, prairie chickens, and upland sandpipers, used these vast stretches of grassland for breeding and foraging. Not surprisingly, these are birds that have nearly vanished from Illinois.

The remaining grasslands are greatly fragmented and isolated. Only a few—Goose Lake Prairie, Midewin National Tallgrass Prairie, Prairie Ridge State Natural Area, and Nachusa Grasslands—are large enough to support species needing expansive areas. And not only are most grassland tracts much smaller, the parcels are widely scattered on the landscape.

## *Poor Quality*

Many of today's grasslands, both cool-season and warm-season, are not in optimal condition for wildlife. The problem is twofold: the composition of plant species is undesirable, and what would otherwise be good-quality grassland is poorly managed.

The modern-day idea of what constitutes a grassland is quite different from the original grassland ecosystem. Grasslands historically were diverse vegetative communities, some with two to three hundred different plants in a given location. Today, one or two grasses, and maybe a couple of broadleaf plants—usually legumes—make up the entire grassland community in many areas. These near-monoculture grasslands typify our pastures and hay fields as well as our ball diamonds, yards, and parks. Some wildlife that evolved with and depend on diverse grassy and broadleaf vegetation struggle to meet their needs.

Some grasslands with only three or four plants can have value to wildlife, but it depends on the plant species. For example, a brome and alfalfa field has high value as nest cover to several species, such as dickcissels, ring-necked pheasants, and meadowlarks, but fescue is universally poor habitat for Illinois wildlife. Unfortunately, it is one of our most widely planted grasses.

Another factor affecting grassland quality is the invasion by aggressive exotic plants. Crown vetch, for example, a legume widely

**Table 3.2 Native Prairie Grasslands**

| Prairie type | Geographic location | Characteristic plants | Secondary plants | Characteristic features | Characteristic animals* | Special management |
|---|---|---|---|---|---|---|
| Loess hill prairie | Along major river bluffs in western and central counties | Little bluestem<br>Sideoats grama<br>Purple prairie clover<br>Scurf pea | Rigid goldenrod<br>Showy goldenrod<br>Leadplant<br>Silky-leaved aster | Loess: calcareous, wind-blown silt. Sites have west or southwest aspect | Glass lizard<br>Surphid fly<br>*Great spangled fritillary*<br>Cone-headed grasshopper | Prescribed burning to control woody invasion. Control of Kentucky bluegrass and sweet clover is needed. |
| Gravel prairie | Gravel deposits in northeast and central counties | Little bluestem<br>Sideoats grama<br>Small-flowered scurf pea | Big bluestem<br>Indian grass<br>Woolly milkweed | Calcareous gravel deposits; many have been mined | Dickcissel<br>*Plains garter snake*<br>Prairie king snake | Prescribed burning |
| Sand prairie | Sand deposits along rivers in central, northwest, and northeast counties | Little bluestem<br>Goat's rue<br>Prickly pear cactus<br>Sand love grass<br>Sand milkweed | Round-headed bush clover<br>Ohio spiderwort<br>Rough blazing star<br>Erect dayflower<br>Cleft phlox | Soil contains much coarse sand; can be very drouthy | *Pocket gopher*<br>Lark sparrow<br>Ornate box turtle<br>Badger<br>Cetonid beetle<br>Yellow mud turtle | Prescribed burning |
| Dry to mesic prairie | Northern two-thirds of state | Big bluestem<br>Indian grass<br>Rattlesnake master<br>Compass plant | Wild quinine<br>Prairie dropseed<br>Prairie dock<br>Sullivant's milkweed<br>Prairie blazing star | Present on deep, dark soils | Dickcissel<br>Plains garter snake<br>Prairie king snake<br>*Prairie cicada* | Prescribed burning |
| Wet prairie | Northern two-thirds of state, bottomlands of major streams | Prairie cordgrass<br>Blue-joint grass<br>Wild blue iris<br>Water parsnip | Turtle head<br>Blue lobelia<br>Cardinal flower | Water present during winter and spring | *Baltimore checkerspot*<br>Massassauga | Prescribed burning to control woody invasion |
| Dolomite prairie | Rock River hill country Lower Des Plaines and Kankakee rivers | Leafy prairie clover<br>Northern bedstraw<br>Riddell's goldenrod<br>Indian plantain | Little bluestem<br>Indian grass<br>Illinois bundle flower | Shallow soil over dolomite bedrock | *Dickcissel*<br>Plains garter snake<br>Prairie king snake | Prescribed burning |

*Italicized animals are pictured below.

planted in the 1970s to control erosion along roadsides and pond banks, has spread to various grasslands. Its aggressive nature causes it to smother out other plants, essentially forming a monoculture with little value to wildlife.

Other non-native species—bluegrass, for one—may provide good habitat in some situations but may be detrimental to remnant and restored natural prairie communities if they get a foothold. Without proper management, bluegrass, a cool-season species that grows profusely in the spring, may invade and outcompete summer-thriving prairie grasses before they have a chance to grow.

## *Poor Management*

Many grasslands that could provide quality habitat are being managed to the detriment of wildlife, with either too much disturbance or disturbance at the wrong time. A case in point is the mowing of roadsides during the peak nesting period of ground-nesting birds and mammals. Hay fields are also usually cut in the same months, further reducing potential habitat.

Another illustration of poor management is the complete *lack* of disturbance in some grasslands. The original prairie flourished with periodic disturbance, such as fire and grazing. Today's landowners may misinterpret the concept of letting nature take care of itself and thus not do *any* mowing or burning. Grasslands depend on occasional disturbance to remain healthy. The areas left untouched may become too thick and matted, and of less value to wildlife.

Also, many grasslands will eventually succeed to woody cover if no disturbance is applied. This result can be seen on our state's remnant hill prairies. On sites where no prescribed burning has been conducted, eastern redcedar and other woody species have invaded, in some cases entirely replacing the prairie grass communities.

HILL PRAIRIE MAINTAINED BY PRESCRIBED BURNING

## How You Can Help Grassland Wildlife

Illinois landowners can help grassland species in two significant ways:

1. Create new grassland habitat.
2. Protect and properly manage existing and newly created grassland habitat.

If you have an opportunity to create new grassland habitat on your property, you should read "Creating New Grassy Cover" (p. 59) for specifics that will enhance your success, from choosing seed to knowing when to plant. Every reader, whether you are creating new grasslands or maintaining or enhancing existing ones, should review "Protecting and Managing Grasslands" (p. 64) and the upcoming section on management.

## Management Considerations

The factors that affect the suitability of grassy cover for wildlife need to be addressed when you undertake a project to provide grassland habitat. When biologists develop a management strategy or plan for a particular site, they think on two levels. First is the *landscape* perspective—the relationship of a particular habitat patch with the surrounding landscape. Second is management of the *patch* itself. Management on both levels is detailed here. Chapter 2 provides further insight into these concepts.

### Landscape-Level Management

Because wildlife don't recognize property boundaries like humans do, it is important to consider the bigger picture of how a particular field or tract of land works with the surrounding landscape to provide regional habitat. In other words, no field or piece of land exists in a void. The animals and plants in and around the site are affected by and interact with the surrounding landscape. In any grassland creation or management effort, be sure to evaluate the following landscape-level considerations.

*Patch size* and *patch shape*. Several wildlife species need large expanses of grasslands to feed, overwinter, and successfully reproduce. For these species, the larger you make your grassland, the better. Since smaller grassland patches (a few acres in size) are more abundant in Illinois than larger ones (100 acres or more), always strive for a larger grassland when possible.

UPLAND SANDPIPER

Some species, such as Bobwhite quail and field sparrows, are attracted to strips of grass interwoven with other cover types. Grasshopper sparrows and bobolinks, on the other hand, need more continuous grassland and less interspersion. With smaller patches the shape is not too important. With larger patches, where the objective is a single expansive and unbroken habitat, a square or circle provides the least amount of edge per area.

*Connectivity* and *adjacent habitats*. Some species benefit from adjacent woody vegetation to provide nesting or song perches. Other patches in the local area can be useful to wildlife, particularly if they are connected by fencerows, waterways, and the like. Species that need large expanses of grassland, such as the upland sandpiper and vesper sparrow, are helped by having other grassy areas, open habitat, or even cropland nearby.

## *Patch-Level Management*

Once you have evaluated landscape-level considerations, you need to determine the management of the site or field itself. To provide suitable habitat for wildlife, a minimum standard must be met for each of four criteria:

1. Appropriate application of disturbance
2. Plant-species diversity
3. Successional stage
4. Structural components

### Disturbance

Whether natural or created by humans, disturbance plays an essential role in any grassland system. Applied appropriately, it improves the quality of grassy cover for wildlife. But used incorrectly, it can undermine habitat value. The following paragraphs discuss disturbance associated with grasslands and when an application is harmful or should not occur. The appropriate use of disturbance to benefit wildlife is discussed in "Protecting and Managing Grasslands" (p. 64).

*Mowing* and *haying.* Mowing at the wrong times can severely affect wildlife. Spring mowing destroys nests and nesting cover, and late-fall and winter mowing reduce valuable winter cover. Both should be avoided. Timing can also negatively affect plant composition, especially with native warm-season grasses. For example, if you mow prairie grasses in late summer, when they are flowering and setting seed, you can set back their growth severely and eventually eliminate them altogether. If your site is to provide adequate habitat, mow only as prescribed in "Grassland Protection with Delayed Mowing" (p. 65).

Black-eyed Susan in tallgrass prairie remnant

*Grazing.* Like mowing, grazing done at the wrong time can harm both wildlife and the grassland itself. Overgrazing can also virtually eliminate value for wildlife, except for the few generalist species that will use barren ground, such as killdeer and horned larks, which already have plenty of bare-ground habitat in Illinois. Heavy grazing can also increase soil erosion, leading to destruction of aquatic wildlife habitats from sedimentation. Grazing should be allowed only as prescribed in "Grassland Protection with Light Grazing" (p. 68).

*Burning.* Burning is an important management tool, but it should be done at the right time of the year to minimize the destruction of wildlife and their nests and young, and to maintain or increase plant diversity. Burning may destroy valuable winter cover, and if done in too large an area with no other standing vegetation nearby, it can reduce or destroy overwintering insects, decreasing an important part of the food chain. Burning should be conducted only as outlined in "Grassland Protection with Prescribed Burning" (p. 66).

*Tillage.* Light tillage in a grassland can provide results for specific management goals, but done indiscriminately, it destroys valuable wildlife food and cover. Till only as described in "Grassland Protection with Tillage" (p. 68).

*Pesticide use.* Herbicides have limited applications in establishing a grassland. Their use should be targeted to specific plants. Otherwise, herbicides can threaten the integrity of any grassland. Insecticides, too, can greatly disturb plant–animal relationships in a grassland. In general, most pesticide use should be avoided there.

*Other human activity.* Building homes, trails, and roads in grasslands disrupts the continuity of the plant community and creates edge habitat. These activities should usually be limited to the edges of any grassland being managed for wildlife. Trails and fire lanes are acceptable, provided they are less than twenty feet wide and make up no more than 5% of the grassland. Less is preferable when feasible.

### Plant-Species Diversity

The more monotypic your grassland, the less wildlife it will support. An optimum grassland should include several grass species and numerous forbs, including some legumes. The array of plants will attract a wider variety of insects, which in turn will provide greater diversity and abundance of food for birds and animals. There will also be a greater variety of plant foods and seeds throughout the year and a diversity of materials for nest building and cover. No single species of grass or forb should make up more than 80% of the grassland. Better habitat can be created by including more forbs. Remember that the original Illinois prairie had 100 to 300 species at any given location.

SHORT-EARED OWLETS AT NEST

## Successional Stage

Different species need grassy cover at differing successional stages. A few prefer bare ground with little or no vegetation. Some species, such as Bobwhite quail and cottontail rabbits, like fields composed of mostly annual broadleaf plants. But most wildlife thrive in more mature fields dominated by perennial vegetation. The amount of woody vegetation in the grassland also affects usage. Some birds, like the northern harrier, use mature stands of grassland with no woody component, while others, such as the field sparrow, prefer grassy habitat interspersed with scattered woody plants.

Depending on the wildlife management objective for a grassland, disturbance should be applied to provide for a specific successional stage. At a minimum, any grassland should have at least 70% perennial vegetation. If your goal is to manage for animal species that prefer larger grasslands with no trees or shrubs and you have at least thirty to eighty acres of grassland, you should use 100% perennial vegetation. If your goal is to manage for more diversity of wildlife, including species that like some woody cover in the grassland, or if you have smaller tracts of grassland, you can develop a mosaic of parcels that provide a mixture of at least 70% perennial vegetation, with any combination of 0% to 30% bare ground or annual vegetation, and scattered woody cover.

From viceroy butterflies to velvet ants, shooting stars to Silphiums, Don Gardner's prairie is a study in diversity. Don's seven-acre prairie restoration is located southeast of Kempton, in northern Ford County. When his great-grandfather settled nearby land in the 1870s, it was prairie. Dairy cattle were pastured on the land until 1965. While growing up, Don explored along the railroad just west of the property, where he encountered remnant prairie plants such as big bluestem and shooting star. Once he became an adult, his dental career left little time for prairies. Over the years, his unattended ground had become an "old field"—until 1974, when Don began a restoration project. Like his ancestors before him, Don was a pioneer. Prairie restoration was a new adventure, and little information or expertise was available.

Don began gradually, putting in a small plot each year. The plot size was determined by the amount of seed he had collected that year. Don was committed to using only central-Illinois seed, collected within a 125-mile radius of his site, to establish his prairie. For the first three years, weedy annuals were all he could see. It was almost enough to discourage him. But at a prairie conference Don met Dr. Robert Betz of Northeastern Illinois University, and he toured Betz's fledgling restoration, a project that mirrored his own. Returning home, Don inspected his plots more closely and could see little seedlings through the weedy annuals. But the hard work had just begun!

While collecting seeds required the most effort, deciding on a tillage system for his new plots provided a challenge of trial and error. First Don tried moldboard plowing and disking. This method provided a fine seedbed but too much soil disturbance—an explosion of weedy annuals resulted. Don then tried chisel plowing and disking. Finally, he sprayed the area to be planted with Roundup, then lightly harrowed, seeded, harrowed, and rolled. The positive results ended his tillage experimentation.

By 1990, the last of twenty plots had been put in, including two control plots that were never planted but received the same management as the others. The "self-restoration" of these plots was remarkable. Don's maintenance activities include late-winter burning, removal

### Structural Components

Structural components, or attributes, like foliage height and density, dictate a grassland's usage by wildlife species. Ring-necked pheasants, for example, prefer thicker grasses such as brome, whereas Bobwhite quail like to nest in old fields with numerous broadleaf plants. Some species, such as the sedge wren, prefer taller grasses, while others, like the vesper sparrow, need short vegetation. Another grassland structural component is litter (dead plant material) from the previous year's vegetation. Some species, like the Henslow's sparrow and many of our native voles, prefer litter for nest building; others, such as the upland sandpiper, need grassy cover without a build-up of thatch. At a minimum, any grassland of more than twenty acres should be divided into ten-acre units, with each unit receiving disturbance only every two or three years. This will provide ample litter on a portion of the grassland while providing a thinner cover on the disturbed portion.

SEDGE WREN

## *Creating New Grassy Cover*

For landowners who hope to see relatively quick results for their efforts, creating grasslands can be satisfying because the vegetation establishes in one to three years. But proper planning is essential to obtain the most diverse, mature grassland in the least amount of time.

of certain exotic species such as sweet clover, and an early-fall combining to harvest seeds of prairie plants.

Don's prairie has 138 prairie plants native to central Illinois, almost double the 71 before the restoration. In September 1995, Don's prairie qualified for and became included in the Illinois Natural Areas Inventory as a Category V (natural community restoration site) prairie.

Since Don's undertaking, red-tailed hawks, kestrels, dickcissels, meadowlarks, pheasants, and rabbits have increased. To Don, the wildlife is just an added benefit. Would he do it again? "Yes" is the quick reply. "To satisfy my intent, I could have had a garden plot, but I had a curiosity. I wanted more than just a collection of plants—I wanted the whole prairie concept. The prairie is constantly changing. It is a study of succession. Yes, you can have prairie gentians, but not before black-eyed Susans. I have gained an appreciation for the prairie—its ruggedness and its ability to survive."

*Susan L. Post*

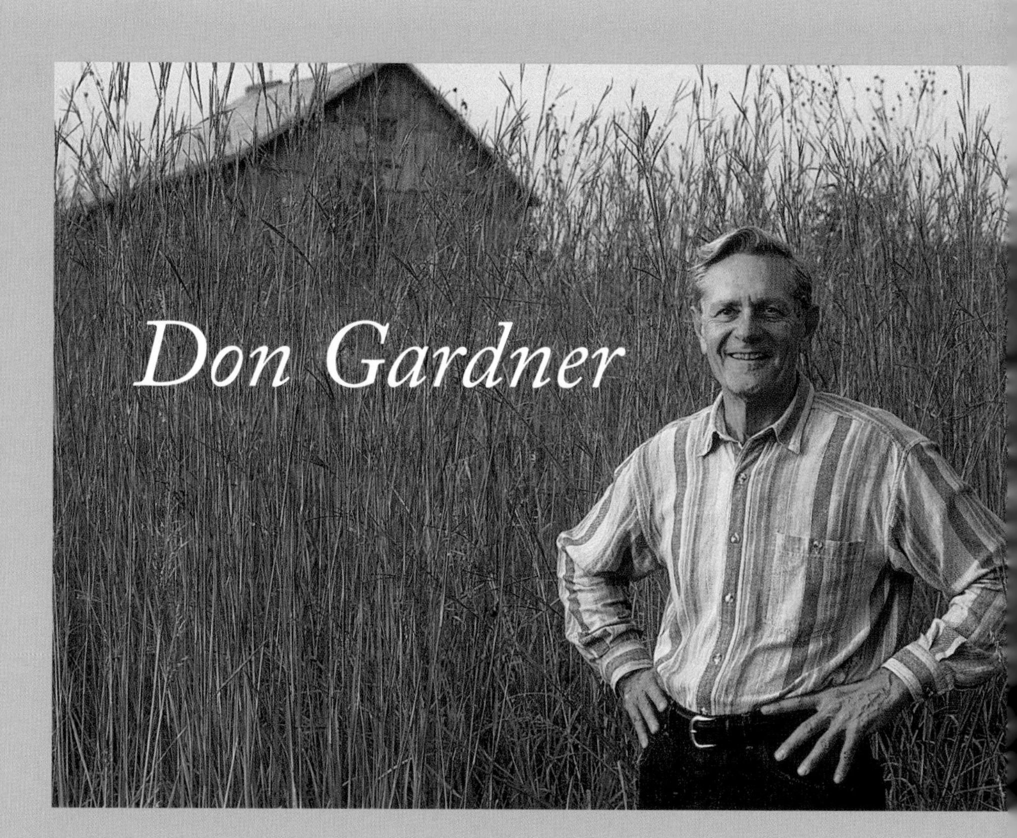

COOL-SEASON GRASSLAND

**Table 3.3 Grassland Plant Combinations Primarily for Upland Wildlife Habitat**

| Species | Pounds per acre | Best sites | Preferred by these wildlife species |
|---|---|---|---|
| *Cool-season grasses and legumes for food and cover* | | | |
| **Redtop**<br>**Timothy**<br>**Korean lespedeza**[a] | 1<br>1<br>5 | Well-drained (mesic); droughty (dry) | Bobwhite quail, for nesting cover |
| **Kentucky bluegrass**<br>**Timothy**<br>**Ladino clover** | 1<br>1<br>1/2 | Well-drained (mesic); poorly drained (wet) | Rabbits, for nesting cover |
| **Smooth brome**<br>**Alfalfa** | 3<br>6 | Well-drained (mesic); poorly drained (wet) | Pheasants, for nesting cover |
| **Kentucky bluegrass**<br>**Timothy**<br>**Ladino clover** | 2<br>1<br>1/2 | Well-drained (mesic); poorly drained (wet) | Bobwhite quail; rabbits, for nesting cover |
| **Ladino clover**<br>**Orchardgrass** | 1/2<br>2 | Droughty (dry) | Rabbits |
| **Redtop**<br>**Timothy**<br>**Alsike clover**<br>**Ladino clover** | 1<br>1<br>3<br>1/2 | Poorly drained (wet) | Bobwhite quail; rabbits |
| *Perennial plants for food plots* | | | |
| **Alfalfa** | 6 | Droughty (dry); well-drained (mesic) | Deer |
| **Kentucky bluegrass**<br>**Ladino clover** | 2<br>1/2 | Well-drained (mesic) | Rabbits |
| **Ladino clover** | 3 | Well-drained (mesic); poorly drained (wet) | Rabbits, deer |
| *Native warm-season prairie grasses and forbs for cover* | | | |
| **Big bluestem**<br>**Indian grass**<br>**Little bluestem**<br>**Native forb mix** | 1<br>1<br>2<br>1 | Well-drained (mesic) | Rabbits, pheasants, and turkeys, all for nesting cover |
| **Switchgrass** *or*<br>**Big bluestem** *or*<br>**Indian grass** | 5<br>5<br>5 | Any<br>Well-drained (mesic) | Pheasants, for winter cover |
| **Little bluestem**<br>**Sideoats grama**<br>**Native forb mix** | 2<br>1<br>1 | Droughty (dry); sandy | Bobwhite quail, for nesting cover |
| **Little bluestem**<br>**Sand lovegrass**<br>**Partridge pea** | 3<br>1<br>3 | Sandy; droughty (dry) | Bobwhite quail, for nesting cover |
| *Native warm-season prairie grasses and forbs for cover* | | | |
| **Little bluestem**<br>**Sideoats grama**<br>**Native forb mix**[b] | 2<br>2<br>5 | Well-drained (mesic) | Prairie birds, for nesting and brood-rearing cover |
| **Little bluestem**<br>**Indian grass**<br>**Big bluestem**<br>**Native forb mix**[b] | 2<br>1<br>1<br>5 | Well-drained (mesic) | Prairie birds, for nesting and brood-rearing cover |

[a] Use "unhulled" if broadcasting as dormant seeding; use "hulled" when planting into prepared seedbed. Korean lespedeza is being phased out because of its aggressive tendencies but is still used in some mixes.
[b] Consult additional publications on the best forb combinations for individual sites.
Seeding rates given are for drilling. Increase 25% when broadcasting and 50% for winter dormant seeding. Price and purchase seed in pure live seed (PLS) amounts. PLS is viable seed minus "trash." PLS could run as low as 40% of bulk weight in some natives. Obtain introduced grasses and legumes locally. Native grasses and forbs may be harder to find. Try to buy Illinois ecotype seed with native prairie grasses and forbs. Adapted by permission from *Small Landowner Guide*, Illinois Department of Natural Resources, Division of Wildlife Resources.

What type of grassland should you develop? There are three types of grasslands you can create:

- *Cool-season grassy cover*—planting introduced cool-season grasses and legumes
- *Warm-season grassy cover*—planting native warm-season prairie grasses and forbs
- *Old field cover*—tilling a site or allowing a crop field to go fallow, thus letting colonizing broadleafs and grasses establish

The following sections explain how to choose which grassland or combination is right for you and your site and how to establish them.

## What to Plant

Whether you create cool-season or native warm-season grassy cover, there are numerous plant combinations you can use, depending on your objectives and your site conditions. Table 3.3 lists common combinations and planting rates for particular sites, along with wildlife species that prefer each combination. Because cool-season and warm-season plants have such different growing requirements, it is important not to mix plants from the two categories in a planting. Always plant cool-season and warm-season plants separately.

The best combinations for providing optimum wildlife habitat in the native warm-season category are the forb–grass mixtures. Grass-only covers may be used as shown, but including forbs will elevate the value of a grassland to wildlife considerably. There are an infinite number of forbs that may be used; which ones you choose is determined by your site conditions and your budget. Consult publications listed in "Suggested Reading" at the end of the chapter for more information.

The pros and cons of establishing and maintaining native vs. non-native grasslands are summarized in Table 3.4. If you have several acres, one option might be to develop a combination of types, based on site conditions, budget, and management considerations. However, prairie can comprise 100% of your acreage if it contains a good mix of grasses and forbs. Determine your management objectives in considering the proportions of each type of grassy cover. Since cool-season cover is usually less diverse, consider including a small native prairie or an old field as a companion patch on a portion of your land. Since old fields don't provide nesting and winter cover as adequately as cool- and warm-season grasslands, consider complementing a large old field by planting one or the other.

**Table 3.4 Native and Non-native Grasslands Species: Advantages and Disadvantages**

| | Advantages | Disadvantages |
|---|---|---|
| **Native (warm-season) prairie grasses and forbs** | As native to Illinois, more tolerant of climate conditions<br>Provide excellent nesting cover and good winter cover | Many grow too tall (4 to 7 feet) for planting locations<br>Can be expensive<br>Usually require 3 to 4 years for a good stand to develop |
| **Non-native (cool-season) grasses and legumes** | Usually cheaper to plant than native species<br>Establish faster and easier than native species | As non-natives to Illinois, not as adapted to climate extremes<br>Usually flatten under snow, leaving no winter cover<br>Must be rejuvenated every 4 to 8 years to be effective nesting cover |

Establishing native warm-season prairie warrants a special mention. While many of us have experienced planting bluegrass or brome and seeing an almost immediate profusion of seedlings, this is not what happens with native prairie grasses and forbs. Patience is paramount in establishing these native grasslands. One of the characteristics that allowed the prairie to thrive in Illinois' extreme climate is the deep roots of the individual plants. The roots of grasses such as big bluestem and Indian grass often penetrate twelve to fifteen feet into the soil. When a seedling of one of these species emerges, it may remain very small above the ground for the first two years while it does most of its growing below the ground, developing its network of roots. Or an individual plant may develop two or three tall grass blades and a single, stunted seed head the first year and not start to spread laterally until the second or third year. Prairie forbs, too, may not bloom for two to four years after germination.

This should not discourage any landowner from establishing prairie. If your goal is to establish wildlife habitat that contains native prairie grasses and forbs, do it! But it is important to be aware of prairie plants' growing habits. Don't judge your sparse planting after the first year as a failure. Be patient.

## *Where to Plant*

Since more than 60% of the land now known as Illinois was once prairie, if you live in the northern two-thirds of the state there's a good chance your site was historically prairie or savanna. If your land is

LEFT: BADGER DEN IN A SAND PRAIRIE
RIGHT: BADGER

flat and contains the rich, black soil of northern, central, or south-central Illinois, it was probably prairie. While grasslands were primarily an upland community, there was also prairie in many Illinois floodplain areas. For instance, cordgrass marsh as well as prairie composed of mesic species like big bluestem and switchgrass existed on the floodplains of most major Illinois rivers. Some of the sandy soil regions contained prairie. And many of the steep south- or west-facing bluffs along the Mississippi and Illinois rivers contained hill prairie.

Much of the southern fifth of Illinois, especially the Shawnee Hills, was originally forested, as were many bottomland areas and deep ravines throughout the state. So these types of sites wouldn't be particularly suited to grasslands. Some of the hills in western and northwestern Illinois were also largely forested. Refer to chapter 9 to learn about determining a site's historical plant community. Cool-season grasses, warm-season prairie grasses, and old fields will all grow well on historic grassland sites.

## Establishment Methods

To successfully establish cool- and warm-season grasslands, follow this sequence of steps: assure proper soil fertility and pH, prepare a good seedbed, seed at the proper dates, use the proper seeding method, and control aggressive weeds.

*Soil tests* and *fertility.* Consult your phone directory or local farm agency for details on soil testing. Here is a summary of suggestions for grassland establishment: Take one soil sample (about eight inches deep) for every three to five acres. Based on the test results, make these amendments if needed before or during seedbed preparation:

- Add potash to bring the test level to at least 30.
- Add potassium to bring the test level to at least 200.
- Correct pH to at least 6.2.

Nitrogen is not needed for native grasses during the first two years of establishment, and it is actually detrimental because it spurs growth of weeds that compete with the small grass seedlings. If nitrogen is necessary, do not apply it until the third year after planting a native grassland.

*Site preparation.* On non-erodible sites with heavy sod, prepare a good seedbed by plowing or using other deep tillage, then destroy new weed seedlings as needed with tillage or contact herbicides until you are ready to plant. On erodible sites, perform all tillage on the contour and leave sufficient surface residue to protect soil. No-till seeding is also an option; it may be preferable on certain sites, such as those that are highly erodible or ones where no-till cropping was used previously and disturbing the soil through plowing would create an onslaught of annual weeds. Check with your local Soil and Water Conservation District office or Illinois Department of Natural Resources biologist if you need advice on site preparation.

*Seeding dates.* It is essential to follow these seeding guidelines; otherwise the seed won't sprout, or the seedlings that sprout won't survive. Table 3.5 shows the range of seeding dates by species and geographic location.

**Table 3.5 Grassland Seeding Dates for Illinois**

| North | Central | South |
|---|---|---|
| ***Introduced grasses and legumes*** | | |
| Late winter to June 1<br>or<br>August 1 to September 1 | Late winter to May 15<br>or<br>August 1 to September 10 | Late winter to May 15<br>or<br>August 1 to September 20 |
| ***Native warm-season grasses*** | | |
| Early spring to June 15<br>or<br>November 1 to March 1 | Early spring to June 5<br>or<br>November 15 to March 1 | Early spring to June 1<br>or<br>November 15 to March 1 |

*Seeding methods.* Good seed-to-soil contact is essential. Place seed 1/16-inch to 1/4-inch deep with a grassland drill or grain drill with press wheels. A broadcast clover seeder may also be used. If you broadcast the seed, roll or culti-mulch (with the tines up) before and after seeding. It is extremely important not to plant seed too deep.

The fluffy seeds of Indian grass and bluestems usually cannot be planted with a conventional drill. A special prairie grass drill is required, and even then it is sometimes necessary to go over the field several times, with periodic stops to feed the seed through the planting tubes. These seeds can be efficiently broadcast by hand on small areas.

Seed may also be mixed and spread with a carrier such as wheat or oats or a dry fertilizer. When using a fertilizer spreader, set it at half rate and go over the field twice, the second time between the previous tire tracks. If mixed with oats or wheat, the grasses may be seeded through a standard grain drill. Use the minimum amount of "carrier" possible.

Aggressively control wheat or oat competition by repeated mowing during the first weeks of the establishment period. Mow the wheat and oats before they form seed heads.

*Weed control.* Control of competing nondesirable grasses and broadleafs is very important to newly established grasslands. Clip weeds above grass seedlings before they grow one foot tall or threaten to shade the new stand. Use a rotary mower; sickle bar cuttings can smother new seedlings. Rake clippings, and remove them if possible to avoid smothering. For warm-season prairie plantings, mowing beyond June is recommended only when tall weeds such as giant ragweed or goldenrod threaten the planting.

Spot spraying of noxious weeds such as musk thistle and Johnson grass can be effective. Herbicides may be used for other problem plants, but you must be extremely careful not to spray adjacent plants or you may eliminate the very grasses or prairie flowers you are working to establish. If Johnson grass, a non-native warm-season grass, is present on your site, plant only forbs or legumes until the Johnson grass is eradicated—and be aware that this might take two or three years. Burning does not eradicate Johnson grass—in fact, it responds well to fire.

GRASSHOPPER SPARROW AT NEST

With careful planning and conscientious establishment procedures, your grassland should provide excellent habitat within two to four years.

## *Protecting and Managing Grasslands*

Three common methods—mowing, burning, and tilling—exist for maintaining and enhancing grasslands for wildlife. Sometimes grazing can also be used. Each method can be used independently, or they may be used in combination. In addition, your grassland needs to be protected from other disturbances, and the standards for plant-species diversity, successional stage, and structural components need to be met. All of these considerations together form a *practice*. By heeding the collective importance of all criteria in a practice when managing your grassland, you can be assured that you're providing suitable habitat for wildlife.

Remember, complete lack of disturbance is not a healthy alternative for our state's grasslands. At least one of these practices should be used for every existing grassland, depending on the type of grassland, the management objective, and the landowner's capabilities.

With all three practices, the most important factors affecting the disturbance's impact on wildlife are *timing* and *amount*. These are discussed in detail for each practice.

## *Grassland Protection with Delayed Mowing*

Mowing at the appropriate times can control invasion of woody plants in grasslands, but mowing favors perennial grasses. Forbs, legumes, and annual broadleafs, which all produce abundant seeds for wildlife, will decrease. Grass litter also accumulates with mowing. Mowing can be useful, but if possible it is best done in combination with burning and light tillage.

WILD BERGAMOT IN TALLGRASS PRAIRIE REMNANT

Avoid mowing any stand of grass between April 1 and August 1, the prime nesting season for most grassland birds. With cool-season grasses, mowing should also not be done between September 15 and February 15 to assure that adequate winter cover remains. If mowing is used instead of burning for managing warm-season grasses, it should be done only in early spring, from February 15 to April 1.

The exception to these rules is when you are *establishing* grassy cover, when you may have to mow to control undesirable weeds. Then mow as needed with a rotary mower, although generally not after June in native warm-season grass plantings.

Don't mow more than a third of an established stand in any one year. This will provide some undisturbed habitat for wildlife and create a mosaic of different plant heights and densities. Divide the grassland stand into units, and rotate your mowing accordingly. For example, if you have thirty acres, you could divide it into three ten-acre management units, mowing each unit once every three years.

### *Grassland Protection with Prescribed Burning*

Fire can be useful for managing cool-season grasslands and is the preferred tool for managing warm-season grasses. However, defining the specific objective for burning and planning before the burn season are both essential to a successful and safe burn.

A primary benefit of burning is preventing litter build-up (thatch) in grassy cover. With warm-season cover this helps warm up

It's hard to drive by Dennis Frey's Hamilton County farm near Belle Prairie City without taking note. The ninety acres of tall grass—big bluestem and Indian grass—surrounding the family's homesite provide an oasis in the agricultural landscape. Though Dennis is a grain farmer, an early job with the soil conservation service piqued his interest in stewardship. When the district conservationist wanted to establish a prairie in the county, Dennis was on the committee. Their discussions came to mind when it was time to put acreage of his own into the Conservation Reserve Program.

While it appears the grassland was always here, it happened only with a lot of hard work. According to Dennis, the most rewarding part was his first good stand of prairie grass, which didn't appear until three years after he started planting. Dennis began small, putting in only five acres in the first year, and he had to master a different kind of farming. He learned that the prairie grasses needed a fine seed bed, and that after the seed is planted, it should be rolled. Picking the proper date to and dry out the soil faster in the spring and allows the prairie species a longer growing season. These factors are important when you are trying to eliminate competing cool-season invaders that don't thrive in warm, dry conditions. In all types of grassy cover, reducing the thatch level also reduces matting and keeps growing plants more erect. Thinned-out vegetation makes it easier for smaller wildlife species to travel within the cover.

Frequent early-spring burning (March 15 to April 15) can help control woody plants. Burning at the end of this period can also help control and eliminate cool-season problem plants in warm-season

PRESCRIBED BURN OF A TALLGRASS PRAIRIE REMNANT

plant was also important—planting the first or second week in June led to an 80% success rate.

From that first five acres the grassland on the Frey farm has grown to over ninety acres. As Dennis continued to plant, he interspersed wildlife food plots into the area, and even integrated a wetland. The wet areas became a "field of dreams" for waterfowl, attracting geese, teal, and mallards.

Once established, the grasses have been fairly easy to maintain. Different parcels are burned every year, with the whole cycle completed every four years. "When it burns, what an awesome sight," says Dennis. "The flames are twenty to thirty feet high and it sounds like a train. I can't imagine a thousand acres burning with the wind blowing. What would you do?"

DROOPING CONEFLOWER

Asked if he would plant his grassland again, Dennis gives a resounding *yes!* "While I was always interested in wildlife as a kid," he says, "these plantings have given me a greater appreciation of nature. Six coveys of Bobwhite live on the property, and the grassland provides an escape for deer as they bound across the lane to disappear into the grass. Red fox can be seen on the hills, and I even have a beaver dam. Just seeing the wildlife and being able to watch the ducks and birds close-up gives me a great feeling. I also find myself spending more time providing for and watching wildlife than I do hunting."

If he could do it over again, Dennis would plant more prairie forbs and start earlier, but overall he is happy. This is the family property where his father was born and farmed. Now it is Dennis's turn to work their special ground. By his incorporation of a native landscape into a traditional grain farm, Dennis has created a treasured place.

*Susan L. Post*

grass plantings. But it will also favor the grasses and cause forbs, legumes, and annuals to decrease. Late-winter burning (January 15 to March 15) will benefit diversity in the plant stand, favoring forbs, legumes, cool-season grasses, and annuals, but it will not be as effective in controlling woody plants, except cedar and pine. Hand removal or herbicide treatment may be needed to control woody plants where spring burns are not desired.

ELK ONCE GRAZED ILLINOIS GRASSLANDS.

Fall burns may be conducted in prairie stands and may be desired on sites that are usually too wet to burn in the spring. Burns done in October or early November can help control or reduce invasion of woody plants and favor forbs. However, fall burning eliminates valuable winter wildlife cover. It is also not recommended on newly established plantings located on steep slopes. Until the grassland has had time to develop its extensive root system, the site could be prone to erosion if a fall burn is conducted.

As with mowing, never burn more than a third of the established grassland acreage in any one year. This rule is especially important with burning because of the insects that overwinter in grassland plants.

See the "Suggested Reading" resources at the end of this chapter for information on safely conducting prescribed burns.

## *Grassland Protection with Tillage*

Tillage can be used to thin a grassland that has become too thick or to establish or promote broadleaf plant diversity in a grass stand. Light tillage (less than 25% of the tilled plot) prevents litter build-up and increases legumes and annual plants. Light tillage may not prevent woody plant invasion. Hand removal, herbicide treatment, delayed mowing, or prescribed burning may be used when woody plants need control. Heavy tillage (more than 90% of the tilled plot) controls woody plant invasion, creates bare soil areas, and will convert most cool- and warm-season grassy covers to old field cover. Any amount of tillage may open prairie stands to invasion by weeds and is not a preferred disturbance practice in prairies that already have a high plant diversity. For seedings that are only grass, however, tillage may improve plant diversity.

As with mowing and burning, never till more than a third of the established grassland acreage in any one year, and do not initiate tillage in new areas during the nesting season, between April 1 and August 1. Fall tillage is often recommended.

## *Grassland Protection with Light Grazing*

Light grazing may be a compatible disturbance with wildlife cover, but the qualifier "light" cannot be overemphasized. Bison and elk historically grazed on Illinois' prairie grasslands, but the grazing patterns were very random and infrequent at most locations. If you plan to graze a grassland, it should preferably be done outside of the nesting season (April 1 to August 1). However, very light grazing during the nesting season may be done without serious consequences to nesting wildlife. Plant heights of at least twelve inches should be maintained at all times on grazed grasslands. Chapter 6 has recommendations for grassland used primarily for hay and pasture.

## Additional Management Tips for Grasslands

In some cases, existing grasslands can benefit from the addition or introduction of desired plants. This can be achieved in two ways: through interplanting (planting the actual plants) and overseeding (planting seed of desired plants into grassland). A common method of interplanting is to place potted plants or root stock of forbs in sparse areas of an existing prairie. Scattering seed of a desired plant, often a legume or forb, over a newly burned or disked stand of grasses is an example of overseeding.

Overseeding and interplanting are usually best done in conjunction with burning and tilling to help temporarily reduce competition for the newly added plants. For seeding legumes like alfalfa and red clover into cool-season stands, tillage is preferred. Till strips in the existing grassland as recommended previously, and broadcast seed into the tilled strips. You can also seed into cool-season stands after burning if the burn is timed to slightly set the grass stand back and provide a head start to the newly seeded legumes.

Seeding or introducing plants into warm-season stands is best done after burning. Remember that burning doesn't kill existing prairie plants, it merely eliminates the tops; so be sure to not interplant into areas saturated with existing plants.

If you have an existing stand of thick native warm-season grasses with few forbs, light tillage will not harm the overall grass stand and will be necessary to successfully introduce prairie forbs.

## Suggested Reading


*Conducting Prescribed Burns.* W. E. McClain. 2003. Illinois-Indiana Sea Grant College Program, University of Illinois.

*Habitat Establishment, Enhancement, and Management for Forest and Grassland Birds in Illinois.* J. Herkert, R. Szafoni, V. Kleen, and J. E. Schwegman. 1993. Division of Natural Heritage Technical Publication #1, Illinois Department of Natural Resources.

*Illinois Agronomy Handbook (23rd edition).* 2003. Circular 1383, College of Agricultural, Consumer and Environmental Sciences, University of Illinois.

*Prairie Establishment and Landscaping.* W. E. McClain. 1997. Division of Natural Heritage Technical Publication #2, Illinois Department of Natural Resources.

*The Changing Illinois Environment: Critical Trends. Volume 3, Ecological Resources. Summary Report of the Critical Trends Assessment Project.* Illinois Department of Natural Resources and The Nature of Illinois Foundation.

*The Tallgrass Restoration Handbook For Prairies, Savannas, and Woodlands.* Society for Ecological Restoration. Edited by S. Packard and C. F. Mutel. 1997. Island Press, Washington, DC.

*Where the Sky Began: Land of the Tallgrass Prairie.* J. Madson. 1995. Iowa State University Press, Ames.

COYOTE AND VOLE

*Chapter 4*

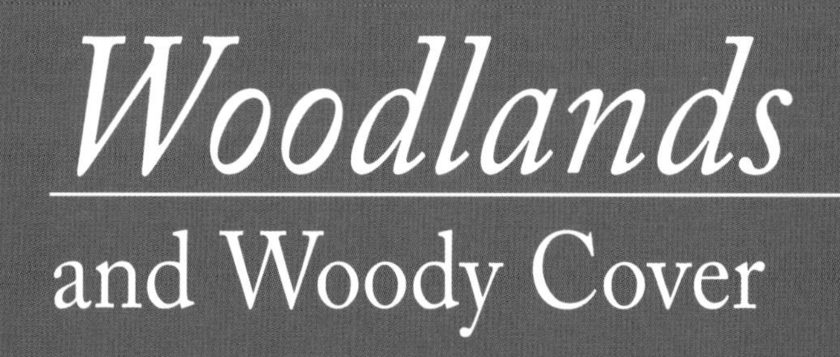

# *Woodlands*

## and Woody Cover

*Listen to the leaves crunching underfoot as you walk along a wooded ridge in the brisk November air. Or maybe you relish the spring forest musical: an orchestra of birds or frogs performing against a pastel back-drop of budding trees and wildflowers. Even the dead silence during a hike through a snow-covered forest can be invigorating. There's nothing like a walk in the woods.*

GREEN TREEFROG
PREVIOUS PAGE: LEFT, WILD TURKEYS; RIGHT, BOBCAT

If you have woodland habitat on your property, you're lucky. If not, why not create some? This chapter will help you make the most of the benefits of new and existing woodland wildlife habitat.

## *What Defines Woody Habitat*

Woody habitat can be described in many ways, based on location, community type, and other factors. For instance, woodlands may be referred to as "upland forest" or "bottomland forest." Another categorization refers to dominant species or groups of species, such as oak–hickory forest, beech–maple forest, and hazelnut thicket. The Illinois Natural Areas Inventory classifies Illinois forests and savannas into twenty-three different community types. To simplify the concept of woody habitat, in this book we address managing and creating all types of woody cover, including forest, savannas, and shrubby cover. For purposes of discussion, we define an area with more than 10% woody canopy coverage as woody habitat and have categorized it into groups.

- *Deciduous woodlands* are closed-canopy woodlands dominated by trees that seasonally lose their leaves, with an understory of shade-tolerant trees, shrubs, and herbs.
- *Savannas* are open-canopy deciduous woodlands that usually contain a moderate or abundant herbaceous layer, often composed of both forest and grassland species.
- *Evergreen groves* are woodlands dominated by coniferous or evergreen trees, those that retain their leaves year-round.
- *Wildlife fencerows* are linear woody cover that may be deciduous or coniferous trees and shrubs or a combination.
- *Shrub thickets* are woody cover composed primarily of closely spaced shrubs or small trees.
- *Shrub borders* are linear woody cover, primarily shrubs and small trees, along the edges of a forest.

To supplement the general information in this chapter, Table 4.1 (p. 75) provides advice for specific types of Illinois forests and other woody habitats.

## *Woodland Habitat Issues in Illinois*

Before European settlement, forests covered about 13.8 million acres, or 38%, of Illinois. Between 1820 and 1920, nearly 80% of our woodlands were eliminated. Since the 1920s, we have seen a reversal of this trend. Woodland habitat has increased in the last seventy years, largely the result of increased planting and natural succession in abandoned pastures. Current estimates indicate some 4.6 million acres of forested land statewide.

Despite the continuing increase in woody habitat, Illinois wildlife that depend on this cover face three problems:

1. The species composition and structure of our forests and other woody cover is changing.
2. The size and interspersion of our woodlands and fencerows is decreasing.
3. The age of our woodlands is declining.

FLOWERING DOGWOOD

## *Changing Species Composition and Structure*

As scientists have learned how ecosystems function, one fact has become clear: fire played a historically important role on much of the Illinois landscape. Many natural communities, including savannas and some forests, were actually fire dependent; they needed periodic fire to maintain community composition and diversity. After settlement, suppression of this natural process played a major role in altering the composition and reducing the diversity of our woodlands. Oak–hickory forest, the dominant community type on Illinois' uplands, is gradually being replaced in many areas—largely because of lack of fire—by shade-tolerant species, such as maple and ash, that are less valuable to wildlife. Research has also shown that the lack of fire and altered river hydrology (human-magnified floods and artificial water levels) have changed the composition of many bottomland forests.

The process of reforestation has also resulted in a different forest composition. Many of these newer successional forests established this century grew on abandoned agricultural lands. At present, these areas are often dominated by less desirable invader species such as honeylocust, box elder, and silver maple and by non-natives such as white mulberry and European buckthorn. Woodlands composed primarily of these species are nearly devoid of hard mast (acorns and nuts) and often lack snags (standing dead trees) and older trees with cavities. They also have a comparatively low value for timber products.

Another type of woody habitat that has nearly disappeared from the Illinois landscape is the open-canopy savanna. Savannas were a different habitat than forests; they were often the transition community between prairie and forest. Savannas had their own wildlife communities, ones that didn't entirely overlap with those of the forest or prairie. For example, bluebirds and flickers preferred savanna as their primary home. Initially, savannas were cleared for cropland because they were easier to cut off than dense forest. Some became forests due to fire supression. Most recently, home builders have been attracted to the remaining savannas because of their idyllic open-shade settings.

Non-native plants have also changed the composition of the understory and forest-floor plant communities, usually to the detriment of wildlife. Many non-native woody and herbaceous plants are very aggressive once they get a foothold. Species such as Japanese honeysuckle,

PRESCRIBED BURN IN AN OAK–HICKORY FOREST

Table 4.1 Deciduous Woodland Restoration Guide

| Woodland types | Geographic location | Characteristic canopy trees | Characteristic understory shrubs | Herbaceous understory | Characteristic features | Special management |
|---|---|---|---|---|---|---|
| **Dry upland forest** | Ridgetops statewide | Post oak<br>Black oak<br>White oak<br>Shagbark hickory | Juneberry | Poverty grass<br>Pussy toes<br>Bird's foot violet | Occur on dry, excessively drained soils | Periodic prescribed burning |
| **Mesic upland forest** | Moist, level to sloping land statewide | Red oak<br>White oak<br>Basswood<br>Sugar maple | Black haw<br>Paw paw<br>Bladdernut<br>Blue beech<br>Flowering dogwood | Mayapple<br>Spring beauty<br>Dutchmen's breeches<br>Wild geranium | Soils have moderately high available moisture | May have to control mesic species and garlic mustard |
| **Bottomland forest** | Along streams statewide | Silver maple<br>Green ash<br>Hackberry<br>Sycamore<br>American elm<br>Red elm<br>Black walnut | Elderberry<br>Spice bush<br>Deciduous holly | Canada nettle<br>False nettle<br>Ontario aster<br>Virginia wild rye | Subject to flooding in spring | May have to control exotic plants |
| **Sand forest** | Sand deposits along rivers | Black oak<br>Mockernut hickory<br>Blackjack oak<br>Black hickory | Prickly ash | Cream false indigo<br>Ohio spiderwort | These forests are very drouthy, and trees are often stunted | Periodic prescribed fire will maintain community; garlic mustard control may be necessary |
| **Sand savanna** | Sand deposits along major rivers in northern half of state | Black oak<br>Mockernut hickory | None | Spotted mint<br>Sand reed grass<br>Cleft phlox<br>Little bluestem<br>Pennsylvania sedge | Occur on dune and swale topography | Prescribed burning to maintain open character |
| **Deep soil savanna** | Northern two-thirds of state | Bur oak<br>White oak<br>Black oak | Hazelnut<br>Prairie willow<br>New Jersey tea | Wild quinine<br>Starry campion<br>Indian grass<br>Big bluestem<br>Golden alexander | Large, open-grown trees with low-hanging limbs | Prescribed burning to maintain open character |
| **Southern flatwoods** | Southern third of state | Post oak<br>Blackjack oak<br>Pin oak<br>Swamp white oak | None | Poverty grass<br>Wood reed<br>Wild quinine | Underlain by hard pan; very wet in spring | Prescribed burning in fall |

CHESTNUT-SIDED WARBLER FEEDING NESTLINGS

garlic mustard, and the bush honeysuckles (Tartarian and Amur) invade forests, smothering or shading out nearly all other plants, including woodland wildflowers, shrubs, and small trees. Climbing euonymus can climb and smother even large trees. The loss of diversity of native plants reduces the variety of natural nesting sites and food, and there is evidence pointing to a corresponding reduction in the diversity of insects, an important component of the food chain.

Exotic plants often gain a foothold in forested areas that have had some human disturbance, such as grazing or logging. And lack of fire may allow the plants to spread; some of these exotics cannot withstand burning like many fire-adapted native plants.

## *Changing Size and Interspersion*

Although forest acreage is increasing, Illinois has about 9.5 million fewer acres of woods today than in 1820. The reduced acreage has spawned other problems, primarily a change in the size and connectedness of the remaining woodlands. Our existing forest tracts are smaller and, for the most part, more isolated from each other. Many woodland wildlife species, including some of our most colorful songbirds, cannot reproduce successfully in smaller forests. And smaller animals, such as salamanders, mice, and lizards, that cannot travel long distances eventually disappear if their habitat is too small and there are no nearby populations to bring in "fresh" genes.

Woody fencerows and windbreaks, which have played an important role in linking fragmented woodlands, have decreased significantly since the 1970s. Agricultural policies that promoted intensive farming were largely responsible. These fencerows not only provided corridors from one woodland to another, they also provided valuable habitat for edge-loving species and stopover points for many migratory birds.

## *Changing Age*

Much of the recent increase in forested acreage is due to abandonment of pastures and new planting efforts. Coupled with continued reversion of mature forests to younger growth or early successional forests through logging, the result is a significant amount of young forest.

Forests commonly referred to as "old growth" are in very short supply in Illinois. This type of forest is characterized by a diversity of age classes, including a canopy of very old trees, 125 to 300 years old. To be considered old growth, a forest would have been relatively undisturbed by logging or grazing for one hundred years or longer, with the exception of an occasional tree cut for firewood or perhaps lumber. Large-diameter trees are important to the survival of some wildlife species, such as the pileated woodpecker. The Illinois Natural Areas Inventory found only 11,600 of the state's existing 4.3 million forested acres to be older, relatively undisturbed forest.

Mature forests are still relatively plentiful in Illinois, but this could change if timber demands increase or the sale of mature woodlands for homesites continues at its current pace. Mature forests are those that may have been selectively logged during the last century but currently have recovered and typically support a wide range of tree age classes. Trees 60 to 100 years old are scattered throughout a mature forest.

It can take centuries to develop a truly mature, full-functioning forest ecosystem. The multiple layers that develop support a diversity of wildlife. Yet a sizeable forest ecosystem can be eliminated in a matter of days! Because it is the slowest of all habitat types to re-create, think carefully before converting forest to another habitat type. Remind yourself how long it takes a tree to grow when you're cutting one down.

## *How You Can Help Woodland Wildlife*

Illinois landowners can help wildlife that use woodlands and other woody habitats in two significant ways:

1. Create new forests, fencerows, and other woody cover
2. Protect and properly manage existing and newly created woody habitats

If you have an opportunity to create new woodland on your property, you should read "Creating New Woody Habitat" (p. 87) for specifics that will enhance your success, from choosing species to planting correctly. Every reader, whether you are creating new woodlands or maintaining or enhancing existing ones, should review "Protecting and Managing Woody Cover" (p. 92) as well as "Management Considerations" (p. 78).

OLD-GROWTH FOREST ECOSYSTEMS CONTAIN MULTIPLE LAYERS OF VEGETATION.

WILD GERANIUMS RESPOND TO A WOODLAND BURN.

such as hawthorn, coralberry, and poison ivy. But the rest of the wooded pasture is barren and void of cover, food, and nesting sites for wildlife.

Even light grazing damages forests. If grazing is not so heavy as to eliminate the understory and floor-layer plants, it will cause a shift in species composition, with the plants tolerant of grazing or not palatable to livestock becoming dominant. Diversity is reduced as some plants disappear.

Not only does grazing destroy the current habitat, but the future of the forest is compromised as new seedlings are continually grazed and destroyed. Grazing also negatively affects existing trees. The prolonged trampling by livestock hooves compacts the soil. Tree roots become exposed and are physically damaged or left susceptible to disease. Hogs will actually root at the bases of trees and can kill a tree in a fairly short time. Studies show that the nutritional benefit to livestock is much lower in a mature wooded pasture than in a grassland. The bottom line is that grazing can damage woodlands, provides inferior forage, and adversely affects wildlife habitat.

*Burning.* Many woodlands can benefit immensely from prescribed fire, but indiscriminate burning can severely damage forest integrity. Burning should be conducted only as detailed in "Woodland Protection with Prescribed Burning" (p. 96).

*Mowing.* Like grazing, mowing can destroy the understory. It should not be done except on a limited basis for specific management needs, such as controlling a brushy understory to promote wildflower growth, to control certain exotic problem species, and to maintain trails, access routes, or fire lanes. Mowing should not be done to promote the look of a city park or to keep a forest "cleaned up." Keep your forest a diverse wildlife haven.

*Cutting.* Cutting and wildlife management may be compatible, but only when done according to a specific plan. Random cutting can result in the removal of valuable trees and can destroy forest integrity. Cutting should be done only as detailed in "Woodland Protection with Timber Management" (p. 94) and "Woodland Protection with Selective Thinning" (p. 92).

*Development.* Moving to "the country" is an increasingly popular choice of people seeking to own a piece of nature and to escape the hassles of urban life. Development poses one of today's most serious threats to our state's remaining woodlands. Taking appropriate measures will minimize the impact of this trend.

If you're building a new home on property with woods, build in or near existing openings (the road, cropland, or grassland) rather than removing trees and fragmenting the forest for your homesite. Also plan your yard, garden, or orchard in or near these openings and near your home to leave a larger core of woodland intact.

If you must build in the woods, insist that contractors protect mature trees around your home, and have as little mowed yard as possible. This approach will provide some continuity with the forest and give your home shade. Avoid driving heavy equipment or piling dirt within twenty-five feet of trees you are keeping. Although mowing woodland around homesites may provide an appealing parklike setting, such disturbance should be avoided if woodland wildlife are desirable. It is cheaper, easier, and better for wildlife if humans keep rural woods as natural habitat.

If you're planning to construct trails or fire lanes in the woods, make them no wider than twenty feet, comprising no more than 5% of the woodland. Trails narrower than six feet are preferable. Minimize erosion by building trails across slopes instead of up and down them.

Try to avoid any tree cutting at your building site from April through July, which is the primary nesting season for most Illinois birds. When cutting in winter, check any trees slated for removal for wintering bats, squirrels, and other animals.

Be aware of non-native flowers and bushes that are known invaders of Illinois forest habitats, and don't plant species such as Amur and Tartarian honeysuckles.

### Plant-Species Diversity and Mast Production

In general, the more monotypic your woody habitat, the less wildlife it can support. Plant-species diversity includes trees, shrubs, and herbaceous plants. Different types of insects and other invertebrates use

### Don't Mess with the Mess

Some landowners, in a desire to make their woodland look like a park or campground, "clean up" or eliminate most of the dead trees, rotting logs, leaf litter, and even living trees and shrubs of the understory. But a forest is much more than an aggregation of tall trees. Even savanna habitats, which are normally open canopy, contain fallen logs, snags, and leaf litter. When it comes to managing a woody habitat, the best rule of thumb is to allow for a variety of structural features, age classes, and plant species. The exception would be removing exotics such as Amur and Tartarian honeysuckles. This would create a short-term gap in the understory but improve the overall woodland in the long term.

## Table 4.2 Mast Producers

The fruit, called "mast," of most trees provides food for wildlife. Acorns and nuts are hard mast, while berries and other seeds are soft mast. Acorn-producing oaks (the *Quercus* species) can be divided between the black and white oak groups; they are considered separately from hickories (the *Carya* species) and other hard-mast producers.

Trees in the black oak group can generally be identified by their dark bark, which may be smooth to furrowed, and by sharp tips on their leaves. They take two growing seasons to develop acorns, which germinate in the spring.

| | |
|---|---|
| Black oak *(Quercus velutina)* | Red oak *(Quercus rubra)* |
| Blackjack oak *(Quercus marilandica)* | Scarlet oak *(Quercus coccinea)* |
| Cherrybark oak *(Quercus pagodaefolia)* | Shingle oak *(Quercus imbricaria)* |
| Hill's oak *(Quercus ellipsoidalis)* | Shumard's oak *(Quercus shumardii)* |
| Nuttall's oak *(Quercus nuttallii)* | Spanish oak *(Quercus falcata)* |
| Pin oak *(Quercus palustris)* | Willow oak *(Quercus phellos)* |

Trees in the white oak group generally have pale gray bark, often in relatively broad slabs or thick and corky ridges. Their leaves have rounded lobes, and their acorns germinate in the fall.

| | |
|---|---|
| Bur oak *(Quercus macrocarpa)* | Rock chestnut oak *(Quercus prinus)* |
| Chinkapin oak *(Quercus muehlenbergii)* | Swamp chestnut oak *(Quercus michauxii)* |
| Overcup oak *(Quercus lyrata)* | Swamp white oak *(Quercus bicolor)* |
| Post oak *(Quercus stellata)* | White oak *(Quercus alba)* |

Because oaks flower early in the spring, just as their leaves are unfolding, they are vulnerable to occasional late freezes that can kill the flowers—and as a result, that season's acorn crop. The delayed second-year development of the black oak group's acorns gives some protection against a total acorn crop failure in a given year. A late freeze that kills the current year's flowers will not kill the developing black oak acorns from the previous year's flowering. Thus two successive killing freezes are necessary to cause a complete acorn crop failure.

Hickories are more dependable mast producers than are any of the oaks. Other nut-producing trees can be grouped with hickories.

| | |
|---|---|
| Beech (*Fagus grandifolia)* | Shellbark hickory *(Carya laciniosa)* |
| Bitternut hickory *(Carya cordiformis)* | Mockernut hickory *(Carya tomentosa)* |
| Black hickory *(Carya texana)* | Pecan *(Carya illinoensis)* |
| Black walnut *(Juglans nigra)* | Pignut hickory *(Carya glabra)* |
| Butternut *(Juglans cinerea)* | Shagbark hickory *(Carya ovata)* |
| Hazelnut* *(Corylus americana)* | Sweet pignut hickory *(Carya ovalis)* |

The soft-mast group includes trees and shrubs that produce berries and other seeds with wildlife food value.

| | | | |
|---|---|---|---|
| Ash | Deciduous holly | Hackberry | Persimmon |
| Blackberry | Dogwood | Hawthorn | Plum |
| Cherry | Eastern redcedar | Maple | Serviceberry |
| Chokeberry | Elderberry | Mulberry | Sumac |
| Crabapple | Elm | Pawpaw | Viburnum |

Browsing white-tailed deer

Black oak–cedar upland forest

different plant species, and different plants produce assorted fruits and nuts at various times. A larger assortment of plants provides a more varied and dependable food supply throughout the year.

No single tree species should make up more than 90% of any woodland. Despite some exceptions in nature, there are general guidelines on diversity. For minimally acceptable wildlife habitat, at least ten species of trees and shrubs need to be present, with no one occupying more than half of the canopy cover. At least two species should be represented from each of the black oak, white oak, hickory, and soft mast groups. (See Table 4.2 for details on these groups.) In addition, at least ten seasonal forest floor (herbaceous) plants should be present. High-quality woods often contain dozens of herbaceous plants.

In addition to ensuring plant-species diversity in any woodland, you need to provide sufficient mast-producing trees; mast is one of the most important food sources for many wildlife species. Timber harvest and selective thinning can greatly alter how much mast is available in a woodland by changing the tree species, age class, and canopy composition. Before undertaking a timber harvest or thinning project, consider consulting with a forester or biologist to ensure that there will continue to be adequate mast production for wildlife after cutting.

## Successional Stage or Age Class

Forests and woody areas of different ages will attract different species of wildlife. To evaluate your current habitat and define your goals, you need to know the different successional stages of woody habitat. Tree diameter is measured and expressed as "dbh," or diameter at breast height, to help create consistency in the way trees are measured.

*Shrub/sapling* is the early successional stage of a forest; it is often a field filled with shrubs and young tree seedlings. The seedlings are usually numerous at this point, all intensely competing for moisture, light, and nutrients. Of the hundreds that start out, perhaps ten to twenty will survive to dominate an acre of woodland.

*Pole timber* stands with trees three to nine inches dbh are often dense, with a heavily shaded understory. If all the trees were planted at the same time or were established concurrently when land was abandoned, the trees in this stand will compete intensely for moisture and light.

*Mature timber* typically contains multiple layers of trees from the ground up to the canopy. Mature trees sixteen inches and more in diameter are scattered about in good proportion, and there is a full complement of seedlings, saplings, shrubs, and pole-size timber. Mast production should be at its peak in this stand. A few trees dominate the canopy, but they are not extremely old.

*Old growth* contains trees more than 100 years old plus seedlings, saplings, shrubs, and pole-size timber. Also present are a full measure of dens, snags, rotting logs, and abundant mast production to support a variety of wildlife.

SYCAMORES DOMINATE THE CANOPY OF A BOTTOMLAND FOREST.

If your objective is a diverse mosaic of forest in different successional stages, consider having 5% to 20% of the land as brushy habitat, pole timber, or both, with mature woodland and old growth each composing 30% to 60% of the site. If you are striving for large patches of a fairly uniform habitat, mature woodland or old growth can compose 100% of the site.

### Structural Components

Dens and cavities, snags, logs, vines, leaf litter, and sometimes rocks or rocky outcrops are important structural components in forests and fencerows. Dozens of vertebrate species of Illinois wildlife are known to use hollow logs or tree cavities during some part of their life. Many use these cavities to bear their young; some use them for winter shelter. Some species prefer hollows in standing trees; others prefer them on the ground. In either location they are valuable forest and fencerow components. Snags offer dens for wildlife, and they harbor a plethora of juicy insects relished by many birds and reptiles. Table 4.3 details some of the wildlife that depend on snags.

To create minimally acceptable habitat, you should include in your woodland both den and snag trees. Some should be ten to twenty inches in diameter, and at least a few should be larger. For ideal habitat, a woodland should contain several den and snag trees per acre. See Table 4.4 for recommended levels of snag and den trees.

Io moth

Remove no more than half of the logs that fall from normal limb loss and tree mortality for minimum wildlife habitat. To create optimal habitat, remove or destroy no more than 10% of the logs bigger than six inches in diameter.

American kestrel at nest cavity

Native fruit-producing vines provide valuable food for numerous animals, offer nesting sites for birds, and create canopy travel lanes for small mammals, reptiles, and amphibians. Retain all native vines, except where they are constricting the trunks of young crop trees selected for future mast production or where they have become dominant and are killing the tops of larger trees. Any non-native vines, such as Japanese honeysuckle and climbing euonymus, should be eradicated.

Rocks and rocky outcrops are locally present in many Illinois counties, particularly in the south, west, and northwest. These rocky areas, which provide valuable habitat for many species of reptiles, amphibians, and mammals, should be kept intact. Rocks for landscaping and other purposes should be obtained from rock cuts, quarries, or other human-made sources.

Leaf litter serves many important functions in forests, shrub thickets, and fencerows. What a brushy fencerow provides in habitat to a rabbit or Bobwhite quail, leaf litter provides to some lizards, toads, and mice. It is a permanent home for many species and a nesting and foraging site for others. Leaf litter also helps regulate soil temperature and retains valuable moisture. And as the leaves decompose they return nutrients to the soil. Except when prescribed burning is appropriate (p. 96), the leaf litter should remain undisturbed.

**Table 4.3 Snag-Dependent Wildlife***

| | Food | Nest | Perch |
|---|---|---|---|
| **Cavity-excavating birds** | | | |
| Common flicker | ✓ | ✓ | ✓ |
| Downy woodpecker | ✓ | ✓ | ✓ |
| Hairy woodpecker | ✓ | ✓ | ✓ |
| Pileated woodpecker | ✓ | ✓ | ✓ |
| Red-bellied woodpecker | ✓ | ✓ | ✓ |
| *Red-headed woodpecker* | ✓ | ✓ | ✓ |
| Yellow-bellied sapsucker | ✓ | ✓ | |
| **Raptors** | | | |
| American kestrel | | ✓ | ✓ |
| Bald eagle | | ✓ | ✓ |
| Barn owl | | ✓ | ✓ |
| Barred owl | | ✓ | ✓ |
| Merlin | | ✓ | ✓ |
| Osprey | | ✓ | ✓ |
| Red-tailed hawk | | ✓ | ✓ |
| Saw-whet owl | | ✓ | ✓ |
| *Eastern screech owl* | | ✓ | ✓ |
| **Open-farm and meadow birds** | | | |
| Bewick's wren | | ✓ | |
| *Eastern bluebird* | | ✓ | ✓ |
| Tree swallow | | ✓ | |
| **Residential-area birds** | | | |
| *Chimney swift* | | ✓ | |
| House wren | | ✓ | |
| Purple martin | | ✓ | ✓ |
| **Water birds** | | | |
| Belted kingfisher | | | ✓ |
| Black-crowned night heron | | ✓ | ✓ |
| Bufflehead | | ✓ | |
| Common goldeneye | | ✓ | |
| Common merganser | | ✓ | ✓ |
| Common or great egret | | ✓ | ✓ |
| Double-crested cormorant | | ✓ | ✓ |
| Great blue heron | | ✓ | ✓ |
| Hooded merganser | | ✓ | ✓ |
| *Wood duck* | | ✓ | ✓ |

*Italicized wildlife are pictured to the right.

| | Food | Nest | Perch |
|---|---|---|---|
| **Woodland birds** | | | |
| Black-capped chickadee | ✓ | ✓ | ✓ |
| Brown creeper | | ✓ | |
| Carolina wren | | ✓ | |
| Great-crested flycatcher | | ✓ | ✓ |
| Prothonotary warbler | | ✓ | |
| Red-breasted nuthatch | ✓ | ✓ | |
| Tufted titmouse | | ✓ | |
| *Turkey vulture* | | ✓ | ✓ |
| White-breasted nuthatch | ✓ | ✓ | |
| Winter wren | | ✓ | |
| **Reptiles and amphibians** | | | |
| *Most salamanders (Redback pictured)* | | ✓ | |
| Treefrogs | | ✓ | |
| **Mammals** | | | |
| Big brown bat | | ✓ | |
| Bobcat | | ✓ | ✓ |
| Deer mouse | ✓ | ✓ | |
| Eastern chipmunk | | | ✓ |
| Eastern pipistrelle bat | | ✓ | |
| Fox squirrel | | ✓ | ✓ |
| Gray fox | | ✓ | |
| Gray squirrel | | ✓ | ✓ |
| Hoary bat | | ✓ | |
| Little brown bat | | ✓ | |
| Mink | | ✓ | |
| Opossum | | ✓ | |
| Raccoon | | ✓ | |
| Red squirrel | | ✓ | ✓ |
| *Red bat* | | ✓ | |
| Silver-haired bat | | ✓ | |
| Southern flying squirrel | | ✓ | ✓ |
| White-footed mouse | | ✓ | |

**Table 4.4 Minimum and Enhanced Levels of Den and Snag Trees**

| | Trees per acre | | |
|---|---|---|---|
| | Less than 10″ dbh | 10-20″ dbh | More than 20″ dbh |
| Snags | | | |
| Minimum level | | | |
| All woodlands | 1 | 2 | – |
| Enhanced level | | | |
| Forest interior | 2 | 4 | – |
| Semi-open habitat | 2 | 4 | – |
| Floodplain forest | 4 | 7 | 1 |
| Den trees | | | |
| Minimum level | | | |
| All woodlands | 1 | 2 | 1 |
| Enhanced level | | | |
| Forest interior | 2 | 4 | 1 |
| Semi-open habitat | 3 | 4 | 3 |
| Floodplain forest | 9 | 14 | 2 |

Definitions: Snag—a dead tree at least six inches dbh at least ten feet tall. Den tree—a live tree with a cavity large enough to shelter wildlife. Forest interior—greater than 80% canopy cover. Semi-open habitat—20% to 80% canopy cover.
Dbh—diameter at 4 1/2 feet, always measured at the uphill side.

## *Creating New Woody Habitat*

Many landowners are reluctant to plant new woodlands or expand existing ones because it takes so long to see the results. But insuring woodlands for the future is of paramount importance, even if those who planted them won't see the final results for years. Proper planning for establishing woodland promotes habitat development at the fastest possible rate. And of course shrub thickets and shrubby fencerow habitat will develop faster than forest habitat.

What type of cover should you establish? Consider the woody cover types mentioned earlier in this chapter.

- For deciduous woodland, plant deciduous (broadleaf) trees and woodland wildflowers with the goal of eventually creating a forest.
- Create a savanna by planting widely spaced deciduous trees or thinning overgrown woodlands and by incorporating appropriate herbaceous plants.
- To establish an evergreen grove, plant evergreen or coniferous trees to create a grove or windbreak.
- To establish a wildlife fencerow, including corridors that connect other woodlands, plant strips of native shrubs or trees (or both) in open lands.
- For a shrub thicket, establish clumps of native shrubs or small trees by planting or thinning to create brushy habitat.
- For a shrub border, establish strips of native shrubs or small trees by planting or thinning at woodland edges to develop brushy wildlife habitat.

Determining what type of woody cover to plant depends on your objectives and the type of site you have. It is useful to determine what type of woody cover may have been present historically on or near your property (see chapter 9).

Your specific objectives will dictate whether you plant a diversity of woody cover types or just one type. For high habitat diversity, you can use a combination of 5% to 20% shrubby cover, 5% to 70% evergreen cover, and 20% to 90% deciduous cover. However, since evergreen-dominated woodland is not a common natural community type in Illinois, we recommend that you limit coniferous plantings to a maximum of one-acre blocks. If you are managing for forest-interior species by trying to reduce forest fragmentation and your planting is "filling in" part of a larger forested tract, plant mostly deciduous trees with only an occasional conifer or shrub.

American Beech in Old-Growth Forest

## *What Species to Plant*

Once you have chosen what type of woody habitat to establish, you need to select species. This again largely depends on your site. Many trees and shrubs tolerate a wide range of conditions and will grow almost anywhere they are planted. But many species, while they may initially grow and look healthy, will develop problems if not planted on an appropriate site. For example, landscape plantings of pin oaks often do poorly. Pin oaks planted in alkaline soil, which includes many of the prairie soils, often suffer from a condition known as "foliar chlorosis." Leaves of affected trees turn yellow, and in a severe case the tree may die. Plant trees on appropriate sites. This promotes landscape health as well as re-creating some semblance of the original Illinois landscape. Refer back to the restoration guide (p. 75) early in this chapter for characteristic species in each woodland community type.

### Spacing and Number of Plants

Once you've decided what species to plant, you can determine spacing. Deciduous trees are typically planted in twelve-by-twelve-foot spacing. To mimic a more relaxed natural aesthetic, you can plant trees in a somewhat random manner. But keep in mind the area needed per tree to supply sunlight, water, and nutrients. Evergreens in a windbreak can be planted up to twenty feet apart; for a grove they can be planted as close as eight feet. Savannas were generally made up of scattered individual trees or clusters of trees; a random pattern with a spacing of twenty to fifty feet will help create savanna-like conditions. Shrubs generally require a four- to six-foot spacing for sufficient density. These spacing recommendations take into account the fact that in any planting a certain number of plants will die before maturity.

Another planting method that can be particularly effective is direct seeding of acorns and other tree seeds. The natural method of regeneration—trees growing from seed without being transplanted from their germination site—can often provide the best results for establishing a forest, savanna, or fencerow. Seedlings germinated on site will often do better than transplants. For information on seed selection, seeding rates, care, and planting, see *Growing Illinois Trees from Seed* in the suggested reading list at the end of this chapter.

### Obtaining Plants

Unless you have a lot of time and money, the most practical way to do large plantings of small trees and shrubs is to use bare-root seedling stock. Potted or balled-and-burlapped trees and shrubs are expensive, can take considerable planting time, and require more care after they're planted.

By having your management plan written or approved by an Illinois Department of Natural Resources (IDNR) biologist or forester, you can obtain no-cost, bare-root seedlings. (However, the seedlings are sometimes limited, so you may not get all the trees or shrubs you need the first year you are ready to plant.) A limited selection of seedlings is also available from many Soil and Water Conservation Districts (SWCD) in spring and autumn. Bare-root stock is also available from mail-order nurseries and some local nurseries. When buying from commercial nurseries, be sure you're buying species native to Illinois.

Acorns, hickory nuts, and other tree seeds may be collected from local trees. This option insures that your source is local and that the resulting trees are suited to your planting location. If you can't collect your seed or are planting a large site, you can also buy tree seed commercially.

If you are planting a small area and prefer to see your seedlings before placing them on the site, try raising your own in pots from acorns or other seed.

### Planting Methods

If you are planting up to 500 tree and shrub seedlings or you have several people to help you plant larger quantities, you can use a tree-planting bar to accomplish the job. Figure 4.1 illustrates its recommended use. For plantings of more than 500, a tree-planting machine may be more efficient. To use a tree-planting machine, you must have seedling-sized bare-root stock.

Tree-planting machines are available for loan or hire from certain IDNR offices, SWCD offices, and local nurseries. A tractor is needed to pull the machine. If you don't have a tractor, the Farm Bureau or Natural Resources Conservation Service can suggest local farmers with tractors for hire. You can also hire a tree-planting contractor to do the entire job.

Acorns and other seeds may be planted by hand with a small spade or shovel. But, again, if large quantities are involved, a planting machine is much more efficient. Acorn planters may be available at some IDNR offices or from local tree-planting contractors.

Gray fox

## Prepare a Site for Planting

To prepare your woodland planting site, eliminate existing vegetation when possible, either by tillage or with herbicide. Treat four-foot-wide strips where seedlings will be planted to provide a jump-start for the young trees and shrubs. Woody plants have difficulty competing with sod. Besides competing vigorously for moisture and nutrients, some grass species are actually "allelopathic," meaning they prevent other plants from surviving near them by producing chemicals that inhibit other species' growth.

In some situations existing vegetation should not be entirely removed from a planting site. If the site is highly erodible, consider removing the existing vegetation only where each tree or shrub will be planted. And don't completely remove vegetation from sites composed of sandy soil or situated on exposed ridgetops or bluffs where winds will continually dry the ground. The loss of moisture through evaporation is actually more of a threat than the competition that existing vegetation poses to new seedlings, and some existing ground cover will help retain moisture on a dry site.

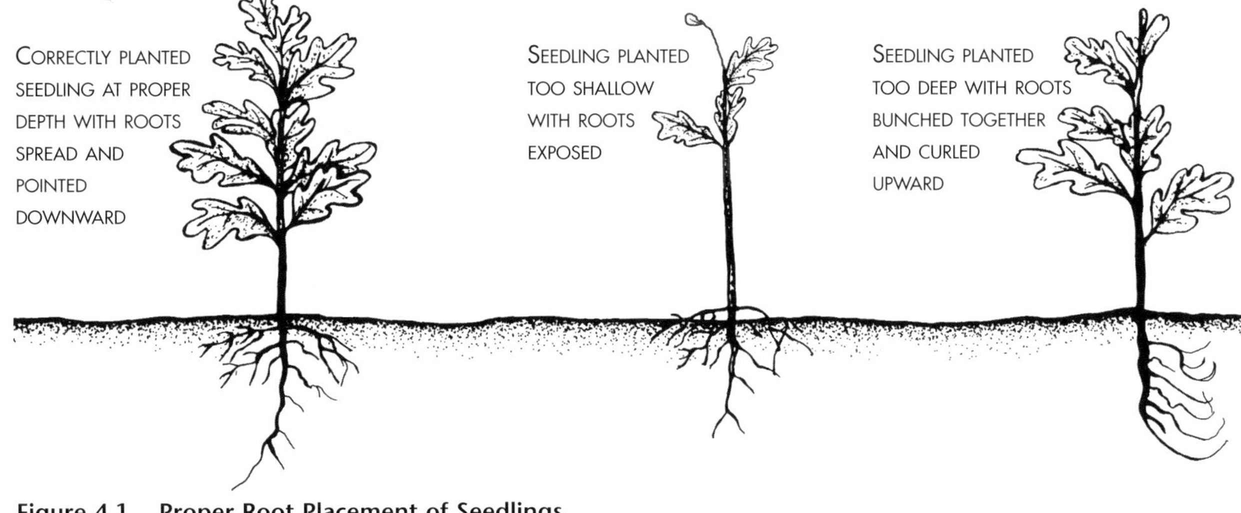


**Figure 4.1 Proper Root Placement of Seedlings**

If you are using an acorn-planting machine, you must till the site before planting. Using the machine in thick sod or weeds will usually clog it, making planting impossible. If the site is highly erodible or dry, try tilling in strips across the slope rather than plowing up the entire field. This practice minimizes erosion.

If you are hand-planting with a tree-planting bar, you can plant into existing ground cover, although it is usually easier in bare soil. Figure 4.1 shows the proper root placement of seedlings. In sandy soil, if the soil is finely tilled and very dry, planting can be difficult because the sand slides back down and fills the slot before you can position the seedling. The easiest solution is to plant after a rain, when moisture will help hold the sandy soil together.

## Seedling Care

If bare-root seedlings arrive in a sealed bag, leave it closed until you are ready to plant. The seedlings can survive for five or six days in these bags as long as they are kept in a cool, shady place.

If you cannot plant the seedlings within a week, you have two alternatives for temporarily protecting them. You may put the unopened bag into a refrigerator, or you may remove the seedlings, separate the bundles, and "heel them in" to soil. Heeling in is accomplished by digging a trench deep enough to completely bury the roots, placing all the plants close to each other, and covering them with a layer of soil. Try to dig the trench in shade and near a water source so you can water the plants. Plants can be refrigerated or heeled in for a few weeks, but seedlings need to be planted by early May at the latest to survive summer heat and dryness.

Never allow the roots of seedlings to dry out or they will soon die. When you start planting, keep the plants in a bucket of water, but not for longer than six hours because the plants may suffocate.

## Protecting New Plantings

Newly planted trees and shrubs face numerous threats. Every care should be taken to insure optimal growing conditions for your new plants.

*Visibility.* Mark seedlings for easy relocation: tie bright tree-marking ribbon on plants, or place wire-stake flags near them (or at least near the plant rows). Leave mowing room between seedling rows and nearby woods, fields, fences, and so on. Don't guess—measure! More plantings fail because of poor weed control and the inability to easily find plants than for any other reasons.

*Weed* and *grass control.* Weeds can quickly rise to heights well above young tree or shrub seedlings, preventing light from penetrating and using up moisture and nutrients. Sod-forming grass can be a threat even more serious than most broadleaf weeds. Particularly threatening are thick grasses like fescue and brome, which not only compete for moisture and nutrients but can actually inhibit growth by "strangling" the seedlings above and below the ground. In addition, fescue is thought to exhibit allelopathic effects on surrounding plants by releasing chemicals that inhibit other plants' establishment and growth.

Mowing, mulching (where practical), and applying herbicides can all provide excellent weed control. Where grasses are thick, elimination by herbicide plus subsequent tillage may be the best choice. And in sandy soils or in any site during a drought year, some shading herbaceous cover around a new tree or shrub seedling can actually offer protection by retaining ground moisture and slowing moisture loss through leaves or needles.

*Moisture.* It is usually impractical to water large plantings of trees or shrubs. But if an unusually long drought occurs and you do have the means to water, identify the healthiest plants and water as many of those as possible.

*Animal damage.* Depending on seedling type and planting location, animals can pose serious threats to stand success. Mice and voles will gnaw on bark, and rabbits and deer will snip off the tops of seedlings. A number of chemical and mechanical repellents have been researched and found to have varying effectiveness. IDNR can provide details on types of repellents. If animal damage is a serious problem, try doing a little less weed control or mowing weeds higher to help hide young seedlings from browsing animals. For a planting of several acres, you may have to accept animal damage as part of the natural mortality of any stand of seedlings.

White-tailed deer

Tubular tree shelters can protect tree and shrub plantings and boost seedlings' growth by collecting moisture and concentrating sunlight. Though expensive, they may be well worth their cost. An alternative that requires less money but more of your preparation time is to remove the tops and bottoms of plastic two-liter soda bottles and place those around new seedlings. Two may be stacked together for more protection. These will offer an early advantage but not the long-term protection afforded by commercial tubular shelters.

## *Protecting and Managing Woody Cover*

Burning and cutting are the two common methods of maintaining and improving woodland habitat for wildlife. They can be used independently or together. Your woodland needs to be protected from all other types of disturbances, and the minimum criteria of plant-species diversity, successional stage or age class, and structural components need to be met. All of these standards collectively constitute a practice. The following practice sections discuss the appropriate use of disturbances, burning, and cutting for woodland harvest and thinning objectives.

### *Woodland Protection with Selective Thinning*

A woodland can benefit from selective thinning when it does not have a desirable mixture of trees, when too many trees are competing for space, or when it contains invasive exotics. Selective thinning may

Behind an Illinois Acres for Wildlife sign, a dozen bluebirds are bunched together, their blue highlighted against the tan of a soybean field ready for harvest. Yet there are no bluebird houses in sight. Nearby a small flock of cedar waxwings gathers. The area is both brushy and open, classic bluebird habitat and a benefit not considered when Steve Lachenmayer began to implement his habitat restoration activities.

Steve has 360 acres of land near Robinson, Illinois. Years ago part of his property was oil land, and three wells still pump tirelessly. His land also contained oak-dominated woodlands and an abandoned homestead accompanied by an overgrown orchard. Of his L-shaped property Steve says, "At the time I didn't realize what I was doing and how it all would fit together. [But] I couldn't have chosen a better piece of ground if I was looking."

Steve loves wildlife, and he also likes to bowhunt. It was his passion for bowhunting white-tailed deer that got him started in habitat restoration, but management for deer was just the tip of what his work has turned out to be. He began in 1992, with the idea of attracting more deer to his property by planting 2-1/2 acres as a clover food plot. It was easy to get started, and he had quick results. By 1995 he planted his first trees. Today he has over 10,000 white pines interspersed with hazelnut, dogwood, and other native shrubs—with established food plots tucked into coves and extending into the woodlands. Only 142 of his 360 acres are now planted to row crops.

With the success of his first food plot, Steve wanted to do more. He contacted his district biologist, and together they came up with a written plan. The biologist thought a variety of food would attract more deer and other wildlife to the area. With this suggestion Steve began to diversify. He thinned the old orchard so the apple and pear trees would again bear and drop fruit. In addition to clover, he planted several varieties of sunflowers, so that the plants would ripen at different times and produce different seed sizes. Turkeys and Bobwhite quail like bigger seeds, while finches prefer smaller. Steve began his woodland management practices by thinning five acres of his woodlands each year. Unwanted trees were girdled and left to decay.

be used for wildlife crop tree management, where desirable trees are protected or released from competition from neighboring, less desirable species. Crop trees may be any species but often include soft-mast species such as persimmon, plum, and crabapple, hard-mast species such as oaks and hickories, and cavity formers such as sycamore, American elm, and post oak. For areas where shrubs are desired, such as shrub thickets, shrub borders, and wildlife fencerows with a shrub component, selective thinning should be used to control succession by removing competing or invading trees.

There are three methods of destroying targeted trees and shrubs: girdling, cutting, and herbiciding. These methods can be used together or separately.

Girdling involves cutting into and through a tree's bark to cut off the transfer of nutrients. This kills the tree standing, which can be desirable because it creates a snag. The tree will topple in time, but the process more closely mimics a natural situation. Girdling is generally used for stems more than ten inches in diameter. It is also the preferred method for thorny species like honeylocust because the thorns decompose with time, resulting in fewer "cleanup" problems.

Simple cutting is another way to eliminate unwanted woody species. Stems smaller than six inches in diameter are better cut than girdled.

Herbiciding can be done independent of girdling or cutting with shrubs and seedling or sapling-size trees, usually as a foliar

Steve considers his white pine plantings to be both the most difficult and most successful part of his restoration. He wanted to provide shelter for the deer, but they kept browsing and rubbing the new seedlings. Steve used Rid-A-Deer, a product that stayed active all winter. While it didn't eliminate the problem, it did keep the deer away long enough that the young pines could gain a foothold.

Steve's hard work is paying off. Plenty of deer stay within a quarter of a mile of the food plots. They are fat and sleek and reach their "full potential." Yet with all of these deer, Steve harvests only one mature buck a year. He has found as he gets older that taking a big buck with a bow becomes less and less important to him. He is content to watch the wildlife from his deer stand—a fox hunting, deer browsing in a food plot, a covey of Bobwhite quail working his land, bluebirds against a gray fall sky. He now wants to do more. As Steve says, "It's about giving back more than you're taking. It goes deeper than hunting."

*Susan L. Post*

*Steve Lachenmayer*

spray. But take extreme caution to prevent drip or drift from killing non-target plants when conducting foliar applications. Where vegetation surrounding and under the undesired tree or shrub is substantial, the risk of herbicide destruction may warrant cutting down the tree rather than applying a foliar spray. Herbicides also often need to be applied to the cut or girdled surfaces of woody plants to prevent resprouting. Avoid cutting during nesting season and midwinter. Fall and late winter are the best times to cut, to avoid disturbing wildlife using the trees for nesting or overwintering. Whenever you cut, try to determine beforehand if larger trees contain bats, squirrels, owls, or other roosting species. If animals are using a tree, monitor their presence and cut once the animals have permanently left.

If snags are lacking in your woodland, consider creating some by girdling or injecting herbicide into trees more than six inches in diameter.

A good use of materials cut down during management activities is to build brushpiles. Chapter 7, "Special Features," gives more details.

When planning your objectives for selective thinning, follow the guidelines in "Management Considerations" (p. 78) on disturbance, plant-species diversity and mast production, successional stage or age class, and structural components.

## *Woodland Protection with Timber Management*

While many people think that logging or cutting for wood products is never compatible with wildlife management, this isn't the case. The activity is harmful only when there isn't enough optimal woodland wildlife habitat or when wildlife considerations have not been carefully woven into the timber-cutting plan. To ensure wildlife needs are met, keep the following points in mind when cutting for lumber or firewood:

Black-oak savanna

- Time your cut carefully. Avoid harvesting during the prime nesting season of April through July. Fall and late winter are the best times to cut, but many wildlife species overwinter in trees. Before harvest, determine if larger trees scheduled to be cut contain bats, squirrels, owls, or other roosting species. If possible, do not cut the tree until any animal using it has left. If harvesting the tree is unavoidable, try to encourage the animal to leave before cutting. However, be aware that disturbing most animals during hibernation or roosting will typically stress them seriously—often causing death.
- Maintain at least seven den trees and snags per acre. Standing dead trees are often of little value as firewood or lumber, but they are of great value to wildlife.
- Do not clearcut. Selective harvest is the method acceptable for many landowners. Clearcutting fragments and alters the infrastructure of the forest, causes the soil temperature to rise, and increases soil erosion in the watershed. However, if your aim is to encourage oak regeneration, selective harvesting of individual trees may not open up the canopy enough to let in sufficient light for oak seedlings to thrive. Group cutting may be in order here; a small group of several trees are cut to allow more light penetration. Contact an IDNR district forester for guidance.
- Leave enough existing trees of all age classes and species to provide a continued variety of habitat and food sources for wildlife. Also be sure to leave some select "parent" trees, especially of highly important species like the oaks, as a future seed source.
- If possible, designate a portion of your forest as "old growth" and leave it permanently unharvested.
- Remember to follow the guidelines in the earlier "Management Considerations" sections (p. 78).

If you're working with a commercial timber harvester, sign a contract that specifies the terms you want and which trees are to be harvested. Incorporate language that ensures minimal damage to remaining trees and the forest floor. Otherwise you may find your woodland severely damaged. Using an IDNR forester is highly recommended. These foresters provide timber harvest advice to landowners at no cost.

GREAT-HORNED OWLETS

## *Woodland Protection with Prescribed Burning*

Current studies and historical research have shown that fire benefits most Illinois forests, especially the oak–hickory communities. This does not contradict what we have been taught by Smokey the Bear—to prevent uncontrolled forest fires. The key here is controlled, or prescribed, fire. Before European settlement, most Illinois landscapes were shaped by periodic fire. But more recently, fire has been suppressed by the elimination of the high-fuel habitats like prairie and by widespread campaigns against "wildfires." This has resulted in ecological changes in our plant communities, including many forests, which need some fire to thrive. The reintroduction of prescribed fire into some forest ecosystems has yielded very positive results.

Savannas in Illinois were completely fire dependent. Without fire, the savanna would often eventually succeed to a closed-canopy forest. Restoring or re-creating a savanna requires periodic burning.

Since fire disturbance is by prescription, it shouldn't be conducted without a purpose. Prescribed fire should be incorporated into an overall land management plan; your objectives will determine the location and timing of the burn. Burning can be used to benefit woodlands in three ways: to control invasion of exotic species, such as bush honeysuckles, garlic mustard, and Japanese honeysuckle; to thin stands of maple and other shade-tolerant species while making conditions favorable for regenerating oak seedlings; and to regenerate herbaceous forest-floor species such as woodland wildflowers. In savannas, fire maintains the herbaceous ground cover, which is often composed of prairie plants. Prairie also needs fire to thrive (see chapter 2).

Prescribed burning requires preplanning and fire management training. For guidance on how to plan and conduct a prescribed burn, see *Conducting Prescribed Burns* in the suggested reading list.

SAVANNA-RELICT WHITE OAK

No more than half of a woodland should be burned at any one time, and burning should be done only from late October through early April to avoid the prime nesting season. In the southern half of the state, wildflowers begin to emerge in March, so burning activities may need to end earlier than April. When planning your objectives for prescribed burning, follow the guidelines in "Management Considerations" (p. 78) on disturbance, plant-species diversity and mast production, successional stage or age class, and structural components.

## *Additional Tips for Managing Woodlands*

Sometimes woodlands can benefit from interplanting—the introduction or addition of desired plants, either alone or in conjunction with selective thinning, timber management, or prescribed burning. Seedling trees, shrubs, or woodland wildflowers and forbs can be interplanted.

A woodland that has been grazed or originated from an abandoned crop field or pasture may contain numerous plant species that do not provide optimal wildlife cover or food. It may also lack the important wildflower component or contain mostly disturbance-tolerant species like poison ivy or exotic, invasive species like garlic mustard and Japanese honeysuckle. A site like this will need to be "opened up" by removal of undesirable trees, shrubs, and herbaceous plants. You can eliminate unwanted vegetation with the methods described earlier. If you are leaving trees in your planting area, be sure not to plant new trees close to the existing ones. Shading and nearby root competition both can hinder the growth of the newly planted trees. Once there is physical space to introduce the new plants, determine what is needed to improve the plant-species diversity of the site. Be sure to use only native species appropriate to the site.

## *Suggested Reading*

*A Long-Range Plan for Illinois Forest Resources.* 1990. Illinois Council on Forestry Development.

*Conducting Prescribed Burns.* W. E. McClain. 2003. Illinois-Indiana Sea Grant College Program, University of Illinois.

*Growing and Propagating Wild Flowers.* H. R. Phillips. 1985. University of North Carolina Press, Chapel Hill.

*Growing Illinois Trees from Seed.* C. A. Hooper and T. W. Curtin. 1983. Circular 1219, College of Agricultural, Consumer and Environmental Sciences, University of Illinois.

*Habitat Establishment, Enhancement, and Management for Forest and Grassland Birds in Illinois.* J. Herkert, R. Szafoni, V. Kleen, and J. E. Schwegman. 1993. Division of Natural Heritage Technical Publication #1, Illinois Department of Natural Resources.

*Planting Hardwood Seedlings.* Illinois Department of Natural Resources.

*The Changing Illinois Environment: Critical Trends. Volume 3: Ecological Resources. Summary Report of the Critical Trends Assessment Project.* 1994. Illinois Department of Natural Resources and The Nature of Illinois Foundation.

*The Tallgrass Restoration Handbook For Prairies, Savannas, and Woodlands.* Society for Ecological Restoration. Edited by S. Packard and C. F. Mutel. 1997. Island Press, Washington, DC.

LUNA MOTH ON YELLOW LADY'S SLIPPER

*Chapter 5*

# *Wetlands*

## and Other Aquatic Habitat

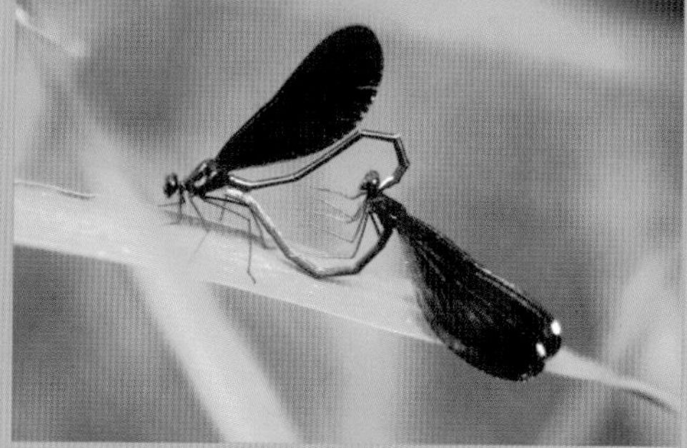

*Water. It's one of the components most desired by landowners trying to improve wildlife habitat. A creek, a pond, or a marsh does add a different dimension to a piece of land. In fact, aquatic habitat can greatly increase the numbers of wildlife attracted to a property. Water not only provides a home for the countless wildlife species dependent on wetlands, but it also attracts many species that just come for a drink, a bath, or something to eat.*

BULLFROG
PREVIOUS PAGE: LEFT, WHITE-TAILED FAWNS IN AMERICAN LOTUS; RIGHT, MATING BLACK-WINGED DAMSEL FLIES

Unfortunately, for centuries some wetlands, especially marshes and swamps, have had a bad image Such areas harbored dreaded diseases during pioneer times. Old World folklore told of the dark mystery of bogs. Who would want such a feature on their land? Water has also been intensely managed in the last 150 years. Illinoisans have extensively modified the environment to make water go where they wanted. Wetlands have been eliminated by drainage and filling. Rivers and streams have been rerouted, deepened, and dammed to suit human purposes.

Anyone with an inkling of interest in nature need only visit a marsh or a swamp on an April morning or evening to hear a remarkable concert of songbirds or frogs. Or you could don waders in November to visit a cattail marsh and see a diversity of waterfowl. See Table 5.1 (p. 102) for information on the attributes of selected wetland wildlife species.

## What Defines Aquatic Habitat

The term *wetland* has historically been defined in many ways, confusing scientists and the public alike. In 1979 the U.S. Fish and Wildlife Service developed a classification system to describe all water habitats in the U.S. The system, classifying wetlands and deepwater habitats, was subsequently adopted by Illinois scientists.

Illinois water habitats are classified in three categories:

- Rivers and streams—aquatic habitats contained within a channel with water that moves permanently or intermittently.
- Lakes and reservoirs—permanent, deepwater impoundments and natural lakes greater than twenty acres in size.
- Ponds, sedge meadows, marshes, forested wetlands, bogs, fens, and other shallow or small habitats—permanent or temporary wetlands partially or completely supporting aquatic plants; bodies of water that are twenty acres or smaller in size.

The terms *wetland* and *aquatic habitat* are used interchangeably in this chapter, with wetland being a generic term for water habitat. Sedge meadows generally are dominated by a mixture of sedge plants. These meadows are not permanently flooded but are located in areas of frequently damp soil conditions. Marshes may have permanent or seasonal water and usually contain both emergent (growing in and above the water) and submergent (growing under the water) herbaceous vegetation. Forested wetlands in Illinois are known by such names as swamps, floodplain forests, and bottoms. Most of our forested wetlands are composed of trees that can survive short, frequent periods of flooding, such as silver maple, cottonwood, and green ash. Southern Illinois, however, has a unique type of forested wetland composed of stands of baldcypress and water tupelo, which are noted for their ability to tolerate long periods of inundation.

While lake and river habitats are discussed in this chapter, the primary focus is smaller aquatic habitats that landowners can create and manage. Also addressed are the habitats flanking rivers and streams (riparian habitat), which landowners can significantly impact. Floodplain forests, which are also defined as wetlands due to their periodic inundation, are discussed briefly, but more specifics can be found in chapter 4.

## *Aquatic Habitat Issues in Illinois*

Just as the face of Illinois grasslands has changed from native prairie to non-native pastures, golf courses, and hay fields, many of our state's original aquatic habitats have been replaced with human-made substitutes. Marshes, temporary spring pools, swamps, and wet meadows have been replaced by fishing ponds, stormwater retention

WOOD DUCKS

Table 5.1 Attributes of Selected Wetland Wildlife

| Species | Preferred foods | Breeding or nesting habitat needs | Winter habitat needs | Additional notes |
|---|---|---|---|---|
| **Wood duck** | Year-round: aquatic and terrestrial invertebrates. Fall and winter: acorns, seeds of bullrush, arrow arum, fruits. | Mature trees (older than 50 years) with large cavities. Uses artificial nest boxes. | Often leaves Illinois and winters in southern U.S. In Illinois. Uses wooded aquatic habitats. | Seldom uses large expanses of open water. Prefers swamps, ponds, and streams with wooded cover. |
| **Canada goose** | Foliage and rootstock of wetland plants and grasses. Also corn, barley, wheat, clover seeds. | Near or in marshes or ponds in undisturbed grassy cover or human-made platforms, tops of muskrat houses. | Open water with access to pastures, grassy areas, and waste grain in harvested crop fields. | A prolonged period of freezing weather combined with a heavy snowfall is usually needed before the subspecies that stop over in Illinois migrate south. |
| **Green-backed heron** | Fish, crayfish, aquatic and terrestrial invertebrates. | Builds a stick nest in trees and shrubs near or overhanging water. Often uses willows. | Usually leaves Illinois in winter. Winters from southern coastal U.S. to northern South America. | Unlike most herons and egrets, a solitary nester rather than colony nester. |
| **Great egret** | Frogs, crayfish, fish, insects. | Nests in colonies or "rookeries" with other species of egrets and herons. | Usually leaves Illinois in winter. Winters from southern U.S. to northern Mexico. | Marshes, shallow ends of ponds, and other shallow wetlands are essential for foraging. |
| **Pied-billed grebe** | Fish, crayfish. Also aquatic invertebrates. | Nests in ponds and marsh with abundant emergent vegetation. | Usually leaves Illinois in winter. Winters from southern U.S. to central America. | Young chicks can frequently be seen hitching rides on a parent's back. |
| **Bullfrog** | Will eat nearly anything, including other frogs, fish, and a variety of invertebrates. | Breeds in nearly any permanent water habitat: ponds, marshes, streams, lakes. Eggs are laid on water surface. | Burrows in mud along edges of ponds and lakes. | Named for bull-like grunting noises made by males in early summer. |
| **Red-eared slider** | Aquatic plants, fish, frogs, variety of aquatic invertebrates. | Eggs are laid on shore in loose dirt or sand. | Burrows in mud in bottom of pond. | Basking areas (logs, platforms, rocks) are essential habitat components. |
| **Mink** | Muskrats, fish, birds, crayfish, mice, voles, some amphibians and insects. | Will use burrows, muskrat houses, or brushpiles, usually near water's edge. | Same as for breeding/denning needs. | Often found along a variety of permanent water habitats, but will range some distance from water. |

basins, and livestock watering holes. Many bodies of water exist in Illinois today, but they differ completely in size, depth, location, and biological character from our original wetlands.

When the pioneers arrived, they found the Illinois landscape dominated by vast expanses of tallgrass prairie interspersed with wet meadows and shallow water marshes. Settlers in southern Illinois and along the major rivers encountered forested wetlands. All of these "wet areas" were viewed as serious impediments to travel, agriculture, and human habitation. The wetlands also harbored disease, since they were reproduction sites for mosquitoes that carried malaria, referred to as "ague" by pioneers.

As Illinois was being settled, ditches and then field tiles were used to drain the wetlands. When possible, wetlands were filled in. Many rivers and streams were also channelized, straightened, or flanked with levees. The levees severed many bottomland swamps and marshes from their natural water supplies, causing them to dry up.

The result of these activities has been a net loss of about 90% of our state's natural wetlands. Human-created lakes and ponds have helped ease the loss of the natural wetlands, but because they have not been made by nature, their biological diversity is, in most cases, greatly diminished from that of the seven million acres of natural wetlands that have vanished from Illinois.

PRAIRIE MARSH

Three factors continue to exert a negative influence on Illinois aquatic habitats and the wildlife that depend on them:

1. The quality of most of our existing aquatic habitats is poor.
2. The size and interspersion of our aquatic habitats are changing.
3. The overall amount of natural aquatic habitats continues to decrease.

## *Poor Quality*

A number of factors contribute to the decreased quality of Illinois wetlands. Unbuffered runoff continues to deliver pollutants (sediment, chemicals, heavy metals, etc.) to many Illinois lakes, ponds, rivers, streams, marshes, and swamps. Pesticides have reduced numbers of plant and insect species, thus lessening food supplies for other wildlife. Sediment from croplands and construction sites is considered the number one threat to all types of wetlands. It muddies the water, creating an environment where many invertebrates cannot survive, thus diminishing life at the lower end of the food chain. Muddy water also makes it difficult for some aquatic predators—fish, amphibians, reptiles, and mammals—to see their prey. Murky conditions that limit sunlight retard or prevent the establishment and growth of many aquatic plants, which are needed both as food and as habitat for aquatic-dwelling animals.

NORTHERN LEOPARD FROG

The negative perceptions of wetlands held by many landowners and the resulting destruction of aquatic plants have also greatly reduced the quality of many of our wetlands. Most biologists define "quality" aquatic habitat as wetlands that support a full complement of interacting plants and animals. Plants provide food for wildlife, surfaces for invertebrates to attach to or lay eggs on, and cover for young fish and amphibians. And ironically, aquatic plants actually help make water clearer—a goal of many landowners. But many pond and lake owners like to see their water free of any aquatic plants, especially to aid fishing. Landowners who have created these fishing ponds have an opporunity to increase wetland diversity by encouraging more aquatic vegetation.

In some parts of Illinois, invasive exotic plant and animal species are destroying intact and productive wetland systems. Aggressive plant species such as the non-native purple loosestrife and the native common reed (phragmites) have created near-monocultures in some marshes, while the invasion of animals such as the grass carp and zebra mussel have reduced native species in various locations.

## *Changing Size and Interspersion*

Fragmentation can be an issue with wetlands as it is with other habitat types. Certain wildlife, such as bitterns and some rails, need extensive acreages of cattail marsh to successfully reproduce. Large expanses of cattail marsh are uncommon in Illinois today.

Wetland complexes forming a mosaic of deep and shallow pools within an area were also once common in Illinois. Such mosaics allowed for great wildlife diversity because there was a more diverse food supply and a variety of places to seek cover and raise young. If one wetland dried up, others nearby would still hold water and

Sora rail

provide the habitat needed by tadpoles, young wood ducks, or herons. Today, individual aquatic habitats are often small and isolated.

Another factor in interspersion of wetlands is their proximity to other quality wildlife habitat. Most borrow-pit lakes along interstate highways, for example, have little wildlife value. They are typically nothing more than bowls of water surrounded by intensively farmed cropland, with little or no adjacent natural vegetation. Many wildlife species need water next to woodlands or grasslands to fulfill their living requirements.

## *Decreasing Amounts*

A surprising number of the few natural wetlands remaining in Illinois still have no permanent protective status and could thus be drained or filled. Federal wetland regulations have slowed the destruction of natural wetlands, but these habitats are still disappearing. Sedimentation and urban sprawl are currently the two primary causes of wetland loss.

Sediment continues to fill many floodplain and upland wetlands. Though sedimentation is a natural process, human activities have greatly increased the amount and frequency of deposition and have disturbed the natural hydrologic cycles that flush sediments out of some wetlands. In a natural setting, some wetlands fill in and disappear while new ones are carved out. In today's landscape, the land is carefully controlled, often preventing the development of new wetlands.

Urban sprawl continues to overrun small wetlands that many people consider worthless. While permits are usually necessary before wetlands can be eliminated, many requests are ultimately

CRICKET FROG

approved. Federal regulations allow this by a process known as mitigation. Rules do require new wetlands to be created to replace any approved for destruction, but it is very difficult to create new wetlands that provide the same benefits natural wetlands offer. Thus, protection of existing wetlands, rather than destruction and replacement through mitigation, should be a high priority for all landowners. In some cases, degradation of natural wetlands is substantial, and active management, including manipulation of vegetation and water levels, is essential to restore suitable habitat for wildlife.

Temporary, ephemeral, and intermittent wetlands are also being lost in Illinois. These terms describe areas that hold shallow water only temporarily or that contain water in some years but not others, depending on precipitation and water-table levels. Throughout our state's history many citizens have questioned the value of these wetlands. These temporary wetlands are some of the most easily destroyed. With the advent of modern drainage methods, thousands of ephemeral wetlands have disappeared with little notice.

These types of wetlands, however, are some of the most important components of the landscape. Semi-aquatic animals such as salamanders, certain frogs, and some toads depend on shallow, fishless water to deposit their egg masses. Historically, these amphibians' life cycles have followed the seasonal rains that formed thousands of temporary spring ponds in Illinois forests and grasslands. Now they face reduced breeding habitat statewide because of the reduction of wetlands, especially the temporary ones.

Ephemeral wetlands are also critical habitat for migrating birds. The shallow waters and associated mudflats provide a banquet of invertebrates for hordes of shorebirds and waterfowl that feed in these habitats. The massive amount of food produced in temporary wetlands and mudflats plays a significant role in the life cycles of these birds, allowing them to complete their journeys to breeding or wintering grounds in a healthy condition. Most Illinois landowners, though, do not recognize the importance of temporary wetlands and thus continue to destroy them.

## *How You Can Help Wetland Wildlife*

Illinois landowners can help wildlife that need aquatic habitats in three significant ways:

1. Create new marshes, swamps, ponds, ephemeral wetlands, and other aquatic habitats.
2. Protect and properly manage

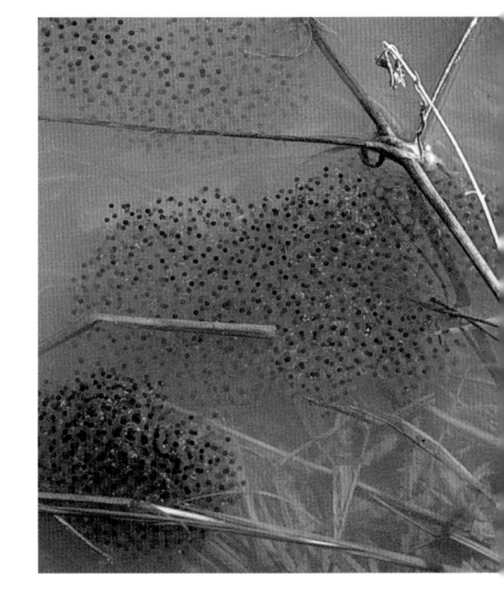

FROG EGGS

existing and newly created aquatic habitats.
3. Restore degraded wetlands.

If you have the opportunity to create new aquatic habitat on your property, read the section on creating new wetland habitat for general information on types of habitats to establish, how to create optimal habitat, and where to get design assistance. Every reader, whether you are creating new habitat or maintaining or enhancing an existing wetland, should read "Protecting and Managing Wetland Habitat" (p. 123). Everyone should also read the upcoming section on management considerations.

## *Management Considerations*

The various factors that affect the suitability of wetlands for wildlife need to be addressed when you undertake a project to provide optimal habitat. When biologists create a management plan for a particular site, they think on two levels. The landscape perspective considers the relationship of a particular habitat site, or "patch," to the surrounding land uses and the regional or statewide landscape. The second perspective is management of the patch itself. Management on both levels is detailed here. Chapter 2 provides further insight into these concepts.

GREAT BLUE HERON

## *Landscape-Level Management*

Since wildlife don't recognize property boundaries like humans do, it is important to consider the bigger picture of how a particular site or tract of land works with the surrounding landscape to provide regional habitat. No field or piece of land exists in a void. The animals and plants in and around the site are affected by and interact with the surrounding landscape. The landscape-level considerations discussed here should always be evaluated in creating and managing wetland habitat.

*Patch size* and *patch shape* must both be considered. Small aquatic habitats can have value. A "frog pond" in a backyard can support a few species of wildlife and does play a role in the landscape. But if you have the opportunity to create a larger wetland, so much the better.

Linear patches of aquatic habitat are usually less desirable than unevenly shaped ones. Irregular edges often create varied habitat along the perimeters of wetlands, making niches in which different types of plants can establish. More diverse plant life can in turn attract and support a larger variety of wildlife. If you have an existing pond or marsh with symmetrical borders, you can take steps to vary the habitat. Creating peninsulas and coves, forming islands, and creating plant mosaics can increase the complexity of your wetland and its attractiveness to wildlife.

AMERICAN TOADS MATING

*Connectivity* and *adjacent habitats* are additional considerations. The proximity of a wetland to other types of habitat can greatly affect its value for wildlife. Can you add a second wetland near an existing one and connect the two with a corridor of grass or woody cover? Can you create a small marsh in a prairie planting, or develop some temporary shallow pools in an existing forest? Look at ways to integrate a proposed or existing wetland with other suitable habitat.

PINTAIL DUCK

Because wetland health is so dependent on the water delivered from its watershed, proper watershed management is an important component of landscape-level management. Land within an aquatic habitat's watershed should be managed to minimize or eliminate the runoff of sediments and chemical pollutants. This requires proper management of agricultural fields, commercial and industrial sites, and residential sites. For example, upland grass buffers help control runoff and add nesting cover.

Most wetland landowners own only part of the land that comprises a watershed. It is important to try to make your concerns known to the other landowners within the watershed and to offer solutions to help prevent erosion and pollution.

## *Patch-Level Management*

Once you have evaluated landscape-level considerations, you need to determine the management of the site or field itself. A wetland community is more than just water and a few cattails. It is an entire system of invertebrates, plants, and vertebrate wildlife in various layers of the habitat, specifically adapted to the aquatic environment's characteristics (calm, deep water, flowing water, etc.). Four management criteria need to be considered on any wetland site. If the minimum standards haven't been met on any one of the four, the wetland will not provide optimal habitat for wildlife. The four criteria are these:

1. Prevention of or appropriate application of disturbance
2. Plant-species diversity
3. Successional stage
4. Structural components

### Disturbance

Some disturbances can be useful management tools when applied appropriately; others are negative any time they occur. Aspects of three disturbances commonly used in managing Illinois wetland habitats (tillage, cutting, and burning) are outlined here; their appropriate application is detailed in the practices section of this chapter. The other disturbances described are common problems in Illinois wetlands, and each must be controlled or eliminated to ensure suitable habitat for wildlife.

Tillage is particularly relevant to temporary and seasonal shallow wetlands. Tillage can have useful application in wetlands, but if a wetland has desirable established aquatic plants, tilling will often damage or destroy them. Tilling should be done only for a specific purpose, such as removing woody seedlings, or to promote the germination of annual moist-soil plants, as prescribed in "Wetland Protection with Moist-Soil Management" (p. 124).

GREAT EGRETS AND CANADA GEESE

White-tailed deer

Cutting to remove invading trees and shrubs in cattail marshes, sedge meadows, and other wetlands can be useful, but indiscriminate removal without a specific objective can disturb a wetland's integrity, especially by eliminating nesting and perching sites used by wildlife. Cut woody vegetation only for specific management objectives.

Burning can be another valuable management tool, but it should be done only at certain times. Untimely burning can destroy vegetation, wildlife nests, and animals themselves. Conduct burning only as recommended in "Wetland Protection with Prescribed Burning" (p. 126).

A number of other disturbances must always be controlled or eliminated. For example, *livestock grazing* and *watering* might seem to be an acceptable disturbance; after all, many types of wildlife, including large-hooved animals like white-tailed deer, come to drink at the water's edge. The problem, however, with domestic animals' using ponds, creeks, and marshes as water sources is the frequency of use and the large numbers of animals that congregate at the site.

Livestock can cause two types of problems: First, the animals' continual movement in and out of the drinking area tramples vegetation on land, leaving the soil bare and free to wash into the wetland. And over time the soil becomes compacted, making it much more difficult to establish new plants. Trampling also destroys aquatic plants either by crushing them outright or by exposing their tubers and roots. The second problem is that the excrement from livestock increases organic matter and nutrient load in the water. The effects are outlined in the upcoming paragraph on pollution.

If an aquatic habitat must be used for livestock watering, limit their access to one small area, just enough that they can get a drink. Even better, fence off the wetland and provide a water source elsewhere on your property.

*Erosion* and *sedimentation* spell the eventual death of any body of water. Imagine scuba diving in a silted lake, where every kick of a fin or move of a hand stirs up a cloud of silt that blocks all visibility. This is the underwater environment that results from sedimentation. Fish, frogs, turtles, crayfish—aquatic inhabitants have a tough time navigating and searching for food in this pea-soup habitat. Although some species do tolerate silty water, research has shown that clear water supports more fish than silty water.

Preventing sedimentation seems like a simple concept—just protect the wetland's watershed. But implementing protective measures is usually complicated. Economic, political, and philosophical forces, rather than solely conservation ones, drive the land-management decisions in watersheds. If an entire watershed lies within an individual landholding, preventing sedimentation may be easier. But most ponds, rivers, marshes, and other wetlands have large watersheds, with intensively used land, making it difficult to eliminate all sources of erosion. Nonetheless, landowners who have wetlands on their property should take every step possible to ensure that erosion is eliminated or substantially reduced. Actions may include installing terraces, practicing contour farming, and minimizing or eliminating tillage on erodible land; growing permanent perennial cover on steep hillsides; protecting riparian habitat (the zone flanking rivers and streams); and installing filter strips.

LONGEARED SUNFISH

*Pollution* has effects on wildlife populations that are yet unknown, but we do know enough to say that many pollutants—certain herbicides and insecticides, heavy metals, sewage effluent—often do real harm to aquatic systems. Organic matter, livestock excrement, or raw sewage, for example, entering the water in large

PRAIRIE STREAM

quantities depletes oxygen as it decomposes, which can result in a fish kill. The simultaneous death of numerous aquatic plants in a pond or marsh can also severely decrease the oxygen level, again possibly resulting in a fish kill. Crop and lawn herbicides can destroy aquatic plant beds and sometimes change water chemistry. Some insecticides eliminate aquatic insects, other invertebrates, and even fish. Some chemicals, including copper sulfate, which is widely used to control algal blooms in ponds, are highly toxic to invertebrates and even fish. And of course fish kills and mass elimination of invertebrates negatively affect birds and other wetland-dependent wildlife higher in the food chain. Pollution control should always be a top priority in wetland management.

*Structural alteration of a wetland* should be undertaken with great care. To minimize detrimental effects from construction involving excavation or filling or removing stream obstructions, consult guidelines available from the Illinois Department of Natural Resources (IDNR) and other agencies. Some construction activities require permits.

*Invasive exotic* and *aggressive native plants* can severely reduce plant species diversity and are hard to eradicate once established. Purple loosestrife and reed canary grass are examples of species that have been promoted at one time or another as valuable wetland plants. Purple loosestrife is of Eurasian origin; the same is suspected

Within the corn-and-soybean farmland of Champaign County exists an oasis—what appears to be a classic prairie pothole. The hole is ringed with cattail, pickerel weed, and arrowhead. In the shadows, a heron waits silently for a meal. Turtles sun on vegetation hummocks. The area is alive with the calls of birds and frogs.

This is no natural pothole, however, but a wetland created by Don Koeberlein. Don and his brother farm 1,500 acres in southern Champaign County. During the late 1980s, a University of Illinois study was comparing nitrate levels in tile drainage water to those in water passing through wetlands. Don figured his land would suit the study well because the topography allowed wetland construction. And he was in the tiling business, so he had all of the equipment for a do-it-yourself project. Within one year, Don created eleven different wetlands, berming areas adjacent to a stream to form small basins. The tile drainage water then would run from the field through the wetlands before it entered the stream.

To most people undertaking such a project, moving acres of soil would have been the hardest part of the job, but Don found "playing in the dirt" easy. For an outdoorsman and hands-on person, doing all the paperwork was the most taxing task. The project required seven permits that took four months to obtain. Before he could start, an archeological survey had to be done. Cost-share money from the U.S. Fish and Wildlife Service and the Farm Service Agency helped with the expenses, and Pheasants Forever provided some of the seed and trees.

Don knows there are many ways to build a wetland and that the results are not always satisfactory. He stresses that a site must be irregular, or have a high edge-to-area ratio, to provide habitat variety. He concedes that he overbuilt his dams and believes they would look better if they were only 2-1/2 feet high. Don merely has to look around, however, to know that his wetlands are successful. "I can't get over how much nature has spontaneously grown, once I provided it the right conditions," says Don.

Don's neighbors were hesitant about his project. They wondered if the created wetlands would affect their production or imply that too of reed canary grass. Both species are very aggressive and form dense stands. Reed canary grass has been widely used for erosion control; however, native alternatives are available and preferred. Species such as common reed, also known as phragmites, which has been widely used in strip mine reclamation, can choke out most other plants to form a monoculture. Phragmites is native to Illinois; however, some suspect that introduced cultivars may have helped create the aggressive plants we see today. In situations where a landowner is trying to promote a diversity of emergent aquatic vegetation, cattails can be aggressive and reduce or eliminate populations of other plants. If you need to revegetate a wetland because a naturally occurring seed source is lacking, use locally adapted native aquatic plants. Avoid using aggressive plants like reed canary grass around the edges of the wetland or on dikes or levees. Reed canary grass should not be used as a filter strip plant; better choices exist, such as switchgrass and big bluestem.

*Invasion of wetlands by exotic species* creates problems. Some aggressive exotic species, such as the common carp, can overrun a lake or pond and degrade the aquatic habitat by rooting in the bottom. This causes silt to suspend in the water, reducing light penetration. Plant growth is reduced, and sight feeders like bass and bluegill are hindered in finding food. Carp can also destroy the nests of other native fish and compete intensely for food resources.

many pesticides were being put on fields. Even his parents were skeptical—until they saw the results. Before Don built his wetlands, nitrate levels were higher in his tile water than in the stream. Today, the drainage water flowing through his constructed wetlands has significantly lower nitrates than those in the stream.

Don even finds himself advising his tiling customers to put in wetlands to intercept tile flow. While he realizes that tile is necessary to drain land, he also knows that it can cause problems. Created wetlands are one way to solve those problems.

What has seeing more wildlife meant to Don? "The beauty I only saw in crops I now see in nature," he replies.

One question Don hears often is "What about mosquitoes?" His answer is simple: "You put nature in balance, and it will work out."

*Susan L. Post*

*Don Koeberlein*

Large populations of certain native species can likewise cause problems in small closed systems, such as ponds. For example, native turtles like the red-eared slider and the painted turtle feed on aquatic plants, and if they're too numerous in a small area, they will completely eliminate these plants. Don't introduce new animals into an existing system unless you are certain of the impact they will have.

### Plant-Species Diversity

The more plant diversity there is in a wetland system, the more types of wildlife it will support. Different types of wetlands contain vastly different plant communities. Shallow marshes, for example, may have large populations of both submerged and emergent herbaceous vegetation. These plants form different micro-habitats; the underwater plants may provide habitat for aquatic insects and tadpoles while the emergents provide habitat for various bird species. Swamps, on the other hand, are dominated by trees and other woody vegetation, although they also contain their own unique assemblage of herbaceous plants. Every aquatic system, including deepwater lakes and ponds, should have as diverse an aquatic plant community as possible. The more diverse a plant community is, the more likely it is to withstand disease, pollution, water fluctuations, and pressure from wildlife feeding.

SLIDERS SUNNING ON A LOG

ARROW ARUM

If you are reconstructing or restoring a wetland, should you seed or plant wetland plants? Opinions vary, and many factors are involved. Are seeds of desired species on the site in an existing seed bank, or is there a wetland nearby that can serve as a natural seed source? Seeding can also occur from far-distant wetlands almost immediately. Seeds can be carried some distance by water. Waterfowl also will bring in some seeds from distant wetlands on their feathers and feet and in their feces. Perhaps a little patience is in order. Wait a couple of years and see what germinates or moves in naturally. In any event, it's a good idea to get the advice of someone who has had experience in wetland reconstruction or restoration in your locality. If the wetland is smaller than five acres, with no other wetland within half a mile of the site, plan a balanced mix of at least five different species of wetland plants from at least three of the following groups: cattails, grasses, sedges, rushes, bullrushes, and broadleaves. Area coverage by emergent plants may range from a minimum 10% to fully 100% of the wetland.

For individual sites larger than five acres or where there is more than one wetland available for management, a portion of a large individual wetland or one wetland in a complex can be composed of a single species, such as cattails or cordgrass, if the objective is to manage for specific wildlife that use these large single-species stands. Any remaining wetland not in the monoculture should be planned with the minimum standards for wetlands smaller than five acres.

For optimal habitat on any size wetland, at least twenty different species of wetland plants from at least three of the groups just listed should be present. In addition, optimal habitat for wildlife diversity is obtained when area coverage by wetland plants is between 35% and 65%, with the remainder being open water. Following this guideline not only creates a good mosaic of habitat from the mixture of emergents and open water but also ensures some permanent water to allow the establishment of submerged plants.

### Successional Stage

While it may not be apparent to the human eye, many wetland systems do age. Lakes and deep ponds advance through successional stages. As sediments and organic matter continue to fill in the bottoms of these aquatic systems, over time they often succeed to shallower habitats. Marshes and other shallow wetlands also go through successional stages. Certain conditions can cause a more rapid change to occur in a wetland. For example, a forested floodplain

Cypress–tupelo swamp

wetland may experience a severe, prolonged flood that kills many of the mature trees. In just a couple of years this wetland will revert to an early successional stage. Likewise, an upland marsh that experiences prolonged drought may succeed from herbaceous plants to woody cover, and without natural or human-directed disturbances it might eventually become a forested wetland habitat. As humans alter natural hydrologic regimes, some aquatic habitats may advance through succession more rapidly. For example, the natural flushing action of flooding may be reduced, thus allowing sediments to fill a wetland more rapidly and causing faster succession.

Rivers and streams also age in a sense. The older some rivers and streams become and the more sediment and organic matter they accumulate, the more they meander, cutting successively wider channels. A broader channel and floodplain is formed, and new wetlands may be formed when side channels are cut off from the main channel.

Aquatic systems of different successional stages attract different types of wildlife. For example, mudflats and shallow wetlands with annual moist-soil plants, which represent an early successional stage, attract a variety of shorebirds. Many of these same birds are not found in mature wetlands with established emergent vegetation; different wading species would appear instead.

Successional stage can be controlled better in some types of wetlands than in others. Humans have tried to control the succession of rivers and streams by keeping them from changing course, but time

Beaver lodge in a southern Illinois slough

has proven that nature often dictates the final outcome. Controlling or retarding succession in other aquatic habitats can be done by a variety of methods, such as reducing the accumulation of sediments and organic matter with proper watershed management. Artificial water-control structures can help manage the successional stage of plant communities in shallow wetlands.

BEAVER

Here are guidelines for providing adequate habitat: One successional stage may be maintained on individual shallow wetlands smaller than five acres. More stages can be provided if desired. Ponds smaller than five acres, however, should have some portion in moist-soil or shallow emergent habitat, in a bay or along a shallow sloping edge. A deepwater pond, without other variation in the terrain, often doesn't provide suitable wetland wildlife habitat.

For wetlands larger than five acres or small wetlands that are part of a larger complex, maintain or create at least two successional stages over the entire wetland area. For example, a portion can be maintained with shallow permanent water with a variety of emergent plants, while the remaining acreage can be used as moist-soil wetland.

## Structural Components

Logs, rocks, snags, and physical features such as islands can all be important structural components in aquatic habitats. Many wetlands contain one or more of these components naturally. If not, you should add some.

Most wetlands that have woody cover in or around them contain partially or fully submerged logs and branches. These logs provide cover, feeding sites, and spawning sites for many vertebrate wildlife species, such as frogs, young fish, and turtles. They also provide valuable substrate to which invertebrates can attach themselves. Some wetlands, such as those along creeks and rivers, also contain rocks, which serve a similar purpose to logs.

Snags and den trees can provide valuable nesting sites for wood ducks, prothonotary warblers, and other cavity nesters. Open tree branches can provide perch sites for fish-eating birds like osprey and kingfishers and insect eaters like swallows and flycatchers.

Islands add great diversity to aquatic habitats. Islands can provide safe nesting sites for many wildlife species. They can be especially important in ponds because they vary the terrain and thus diversify the available habitat.

A greater variety of structural components in a wetland will attract more types of wildlife. Chapter 7 discusses the use of these structural components in more detail. Minimal standards for creating adequate wildlife habitat include the following: Provide or maintain at

GRAHAM'S CRAYFISH SNAKE

BLUE-WINGED TEAL

least five structural features per acre of aquatic habitat. At least three should be logs in or along the edges of the water, and at least one should be a perch site. For optimum habitat, more components (as many as ten per acre) should be present.

## *Creating New Wetland Habitat*

Compared to establishing a mature forest, creating a marsh or pond takes a relatively short time. A few wildlife species, especially birds, may come to use the wetland as soon as it starts holding water. A mature, fully functioning aquatic habitat does take a lot more time to develop, but proper planning will result in more rapid and successful establishment.

What sort of wetland habitat should you establish? Four types can be created by Illinois landowners:

- *Pond.* Establish impoundment to provide wetland habitat for wildlife and water deep enough for fish.
- *Marsh.* Establish shallow water area and herbaceous vegetation to provide water and wetland wildlife habitat.
- *Swamp.* Establish shallow water area and woody vegetation to provide water and wetland wildlife habitat.
- *Ephemeral wetland.* Establish sites to hold temporary or seasonal pools of water or mudflats (or both) for wildlife habitat.

The type of wetland you decide to establish will depend largely on soil type, site, and cost, but your specific wildlife objectives can determine your wetland creation if the physical and financial conditions will allow for the wetland of choice. First we'll discuss the physical factors that dictate a site's potential.

If your soil is permeable, such as sandy or gravelly soil, it's unlikely you can establish any permanent wetland. Constructing a pond or lake requires a natural ravine or depression that can be blocked off with a dam or dike. Simply "digging out" a pond or lake in flat earth is usually prohibitively expensive. But even building a small dam across a ravine can cost tens of thousands of dollars. Every site and situation are different, and contractors' prices vary, so it is advisable to get a preliminary idea of what you want and have a professional from the local Natural Resources Conservation Service (NRCS) look at your site.

A shallow wetland such as a marsh, swamp, or ephemeral wetland can be constructed by scraping out a depression or constructing a dike. The cost will depend on the existing topography and the amount of earth moving involved. In flat areas, a shallow water marsh can often be established by simply breaking a drain tile or plugging up a ditch. This is an inexpensive way to create a wetland, but be sure the site isn't subject to any legal drainage constraints such as restrictions against impeding, diverting, or draining water; check with the local Soil and Water Conservation District (SWCD) first.

Once you've determined what is financially, physically, and legally possible for you and your land, you need to determine your objectives for creating an aquatic habitat. Are you interested in duck hunting? Are you a bird watcher wanting to attract a variety of shorebirds, waterfowl, and wading birds to your property? Or do you want to have a place to attract breeding frogs? Your goals can guide you in deciding what to create. Duck hunters and shorebird enthusiasts will want to create wetlands that they can drain at the proper time to create mudflats for foraging and to allow the germination of food plants for waterfowl. Shallow permanent waters will attract a variety of wading birds, such as herons and egrets, and aerial foragers such as swallows, chimney swifts, and martins. Shallow wetlands will also be used by mallards, blue-winged teal, and Canada geese in summer and by a variety of waterfowl during migration. Wood ducks prefer a wetland with a woody component, which is also attractive to certain woodland salamanders and frogs. Many frogs and salamanders need wetlands with minimal predator pressure for successful reproduction. A shallow wetland might be considered to attract a diversity of amphibians while still providing habitat for waterfowl.

If your land's topography and your budget allow, consider creating a "wetland complex." A mosaic of aquatic habitats in close proximity can be managed individually to create a broader array of habitat conditions. For example, you could have one or two shallow marshes with permanent water and many cattails, another marsh that is seasonally drained to create mudflats or moist-soil plants, and another marsh that is flanked with shrubs or trees. The more diversity you create, the greater the variety of wildlife you'll attract.

LEFT, BULLFROG
RIGHT, EDGE OF A NATURAL GLACIAL LAKE

## *Designing a Wetland*

This book provides only the basics about choosing a wetland design. Unless your project is small enough that you will construct it by hand, you should seek the advice and technical assistance of a professional. Usually, your county's NRCS can help you design an aquatic system. They can advise you on the suitability of your site, the estimated cost, and engineering specifics. It can be risky to hire a contractor to build a dam or move dirt without knowing your site's suitability for the project or the parameters of its watershed (and thus the amount of water that will be delivered to the pond or marsh). Seek professional assistance first.

Whatever type of wetland you develop, try to create varied habitat conditions through a mosaic of uneven surfaces in the wetland bed and around the edges. With a marsh or forested wetland, construct some portions of the site with little depth to allow for frequent mudflat development. Other portions can be designed to promote emergent vegetation such as cattails, bulrush, and arrowheads. With ponds and deepwater habitats, try to construct some variability. Create islands or shallow areas during construction. Design the pond or lake so there are plenty of fingers with shallow bays.

Two feet or less of water that occasionally dries up will generally promote emergent plant growth. Two to four feet is usually needed

CANADA GEESE AT AN ABANDONED STRIP MINE

TURTLEHEAD ARROWHEAD WILD BLUE IRIS WITH MAYFLY AMERICAN LOTUS AMERICAN FEATHERFOIL

for submerged plants to flourish. Creating gentle rather than steep slopes along the edge of the marsh, pond, or swamp will expose mudflats and foster moist-soil plant development during dry summer weather, even without a draw-down device.

Although flooding a forest to create a pond is sometimes done in Illinois, it is not recommended. If a woodland must be destroyed to create a pond, leave some of the soon-to-be-dead trees standing in or at the edge of the pond to provide nesting and roosting cover for wildlife.

## *Types of Plants and Structural Components*

As discussed earlier in this chapter, seeding or planting a reconstructed or restored wetland may not be necessary, and it can be expensive. Ultimately, the landowner will have to decide whether costs are prohibitive. The types of plants you introduce or allow to grow in the wetland will depend on your objectives and on what is native and thus adapted to the area. A balanced combination of plants will provide more habitat diversity. For minimally suitable habitat for wildlife, follow the guidelines in the earlier section on plant-species diversity.

For example, if you have created a three-acre aquatic habitat with a variety of water depths, you could plant a 1/4-acre corner with hardwood species such as pin oak, green ash, swamp white oak, sycamore, and shellbark hickory. You could establish emergent species such as burread, wild irises, arrowheads, and spatterdocks along another one or two acres. Let half an acre seasonally emerge as mudflats and sprout annual plants such as wild millet and smartweeds, and leave the remainder as deeper, open water. This is just one way to design the habitat; infinite combinations are possible.

Many herbaceous and woody plants will establish on their own when a new wetland is constructed. However, some management may be needed, because willows and other less desirable species may invade and keep preferred species from colonizing. If you want to supplement the natural colonizers or create a specific wetland plant community, private nurseries can provide many species, either as rootstock, potted plants, or seed. Be sure to purchase plant materials

Water measurer on salvinia

grown or collected from the local native gene pool. You can collect seed and grow the plants yourself. Always obtain permission from a landowner to collect seed, and never take more than a quarter of the total available seeds from any source annually. Do not dig up and transplant wetland plants unless a landowner is planning to eliminate the plants and has given you permission to remove them.

Table 5.2 lists some of the common wetland plants that are beneficial to wildlife, are attractive, and are readily available from nurseries or by collecting seed.

Structural components are important to some wetland systems; follow the guidelines given earlier in this chapter. Introduce some logs and branches into the pond or marsh. Provide basking areas for turtles, dragonflies, and other species by placing some logs so they are only partially submerged. Submerge other logs and branches at various depths to provide underwater habitat for invertebrates, amphibians, and fish. Rocks may be placed in a similar fashion, some underwater and others exposed. If the wetland is in the open without any trees, install vertical structures as perching areas for a variety of birds, like swallows and kingfishers. Chapter 7 gives additional details on various structural components, such as islands.

Johnny darter

## *Introducing Fish Into an Aquatic Habitat*

Fish management is not covered in this book. Consult a fisheries biologist with the IDNR for more information. Remember, though, if you want to manage for amphibians, consider creating a fishless wetland. Many species, especially salamanders, cannot breed and survive in a pond that contains fish because the fish eat the eggs and young.

## *Alternatives to Creating a Traditional Pond or Marsh*

If a larger system proves financially or physically impossible, there is still the option to make some small "frog ponds." All you need is a shovel and some time (or money, to hire someone to do the work!). Small versions of the wetlands previously discussed can provide excellent wildlife benefits and add diversity to your property's habitat plan.

If your soil type is not suitable for holding water, artificial liners can be used with some success. This solution is usually prohibitively expensive on a large scale but can be reasonable for backyard ponds. The most successful method for well-drained soils is to sandwich heavy-duty, rot-resistant plastic liners between slabs of a material called Agri-fabric, combining water-holding ability with protection

**Table 5.2**

| Wetland types |
|---|
| Open water marsh |
| Cattail marsh |
| Wet meadow/ sedge meadow |
| Cypress/tupelo swa |
| Shrub/scrub wetla |
| Ephemeral wetland |

**tlands Restoration Guide**

| ographic location | Flooding regime | Primary plants | Characteristic features | Animals | Special management |
|---|---|---|---|---|---|
| ewide, but especially thern Illinois | Intermittent to permanent, 2 to 6 feet | Water plaintain, smartweeds, river bulrush, cattails, arrowhead, water lilies | Hydric soils, many plant species | Rails, bitterns, muskrats, | May have to control cattails or glossy buckthorn |
| thern counties along major rivers | Permanent, 6 inches to 2 feet | Cattails, bur reed | Less plant diversity than open-water marsh | Rails, bitterns, muskrats | Control of reed canary grass, phragmites, glossy buckthorn |
| thern counties | Permanent, 6 inches or less | Blue-joint grass, tussock sedge, seedbox sedges, cordgrass | Occurs on peat, muck, or wet sand, saturated soils | Tiger salamander, northern leopard frog, bullfrog | Control of aggressive invaders like glossy buckthorn and reed canary grass |
| thern counties along he and Ohio rivers | Permanent, 6 inches to 2 feet or more | Spider lily, baldcypress, tupelo, Virginia water willow, swamp rose, button bush | Periodic flooding renews water levels | Wood duck, prothonotary warbler, green treefrog, bantam sunfish | Water level control may be necessary |
| ewide along rivers | Permanent, 6 to 12 inches | Button bush, pussy willow, mallow, sandbar willow | Predominant vegetation is shrubs | Variety of waterfowl | Sedimentation is a concern along many rivers |
| ewide | Temporary, 6 inches to 2 feet or more | Smartweeds, soft-stem bulrush, blunt spike rush, cattails, rice-cut grass | Greatly influenced by amount of precipitation | Variety of waterfowl | Maintain land features that create wetland |

from tears. Agri-fabric may be difficult to find; check with your local SWCD for information.

## *Protecting and Managing Wetland Habitat*

Three common wetland management activities are conducted in Illinois to improve or enhance habitat for wildlife: manipulating water levels for moist-soil plant growth or mudflat exposure, timed flooding of forested wetlands (also called "green-tree management"), and prescribed burning of wetland vegetation. Unlike other habitats, wetlands can provide high-quality habitat if they are simply protected and maintained, without any active management. All wetlands should be protected from any disturbance except the three specified here and need to meet the minimum standards outlined in "Management Considerations" for plant-species diversity, successional stage, and structural components. All of these considerations together comprise a practice. If you remember the collective importance of all the criteria in a practice when managing your wetland, you can know you're providing suitable habitat for wildlife.

At a minimum, any existing or newly created pond, lake, marsh, swamp, wet meadow, or other wetland should be protected from detrimental disturbances such as pollution and dredging. If littering has previously occurred, remove the trash, such as empty containers and old tires, to improve the water quality.

Protection of the watershed is key to protecting any aquatic habitat from sedimentation. Avoid forest clearcutting, which significantly increases the runoff and the sediment load to the body of water. Clearcutting along the banks can raise the temperature of stream water, often causing a negative change in the ecosystem. Minimize or eliminate any unnecessary tillage in the watershed. If tilling must be done, use soil conservation techniques such as contour farming and "minimum tillage." Installing filter strips and riparian buffers can also greatly reduce the sediment load and pollution that would otherwise enter the wetland. Filter strips can be any type of grass that will stabilize the soil around the wetland and provide nesting cover. Fescue and reed canary grass are not recommended because they provide little additional value to wildlife and aggressively outcompete other wetland plants. Wetlands and ponds located in the middle of crop fields can greatly benefit from the establishment of filter strips without the sacrifice of much cropland. Filter strips in any location should be at least 50 feet wide; wider is even better. Consult the chapter on grasslands for species and planting specifics on grass filter strips.

Riparian buffers of trees and shrubs can also trap sediments and pollution and can stabilize banks that are eroding. Plant woody species that are suited to bottomland and moist habitats. The buffer strip should be at least 50 feet wide; wider is better. Consult the chapter on woodlands and woody cover for specifics on species and planting.

## *Wetland Protection with Moist-Soil Management*

In some wetlands, artificially manipulating water levels can create a mosaic of wetland types that attract a variety of wildlife. Water levels and vegetation can be managed to produce herbaceous plants, mudflats, and open water.

GREAT EGRETS

To create conditions favorable for moist-soil plants to grow, in the fall gradually flood the wetland with two to ten inches of water, and let remain over winter. Gradually dewater from ice-out until midsummer; this allows spring migrating birds to feed on invertebrates. The plant seeds will germinate during summer, once water is drained. Vary the flooding and dewatering times from year to year to encourage vigor and density of desired plants.

Biologists have found that food production for several wildlife species, including waterfowl, can be enhanced by duplicating the natural cycles of wet and dry periods that occur when lake or stream water levels fluctuate. Typically a marsh's water level is lowered during the late spring or early summer, allowing moist-soil plants to grow in the exposed mudflats. Tubers and seeds provided by the new plants are used by species like the mallard. Such water management encourages plants such as cattail, bulrush, ferns, common reed, and erect sedge. The drawdown also provides areas for birds to feed on invertebrates. In the fall, water levels are allowed to rise again several feet, inundating the mudflats and creating additional areas for migrating waterfowl to rest and feed. The timing, duration, and frequency of drawdowns are all important. Drawing down earlier may encourage certain plant species, while a later drawdown will help completely different ones. For specific information on the benefits of mudflats to wildlife and how to develop and maintain them, see the U.S. Fish and Wildlife Service *Waterfowl Management Handbook* in the suggested reading list.

## *Wetland Protection with Green-Tree Management*

Green-tree management aims to mimic the natural flooding that occurs in some bottomland forests during dormancy. A great variety of wildlife benefit, including migrating waterfowl, birds of prey, bottomland songbirds, and reptiles and amphibians.

MUSKRAT

Flooding should begin gradually in late fall, when trees are fully dormant. Water levels should be maintained at a fairly stable depth of twelve to eighteen inches during winter. Before trees leaf out in the spring, a gradual drawdown should occur, over a period of at least two to three weeks. Try to match the drawdown period with the arrival of migrating waterfowl. Remember, you are trying to mimic what occurs naturally in some bottomland forests.

Stable water levels are desirable in midwinter because suspended nutrients usually peak about three months after flooding. Avoid rapid drawdowns. Gradual drawdowns prevent a high loss of nutrients, result in more stable wildlife habitat, increase invertebrate availability, and promote the long-term health of the system.

Do not flood the woodland every year, and rotate areas so they are flooded only every other year or less. Flooding can alter bottomland plant communities. Pin oaks and swamp white oaks tolerate extended flooding. Other bottomland trees and some wildflowers

**Table 5.3 Trees Tolerant of Extended Flooding**

| | |
|---|---|
| Black willow | Honeylocust |
| Box elder | Pin oak |
| Bur oak | Silver maple |
| Cottonwood | Swamp white oak |
| Green ash | Sycamore |

American broad-banded water snake in flooded bottomland

that inhabit bottomlands, such as bluebells, do not. Table 5.3 lists ten tree species that tolerate extended flooding. If you have questions about this practice, contact the IDNR.

## *Wetland Protection with Prescribed Burning*

Introducing occasional fire during the late fall or early spring to a wetland that is artificially or naturally dry can greatly improve the structure and diversity of the vegetative community, thus improving the wetland for wildlife. Conduct a prescribed burn on emergent, herbaceous aquatic vegetation (cattails, sedges, grasses). Burn no more than a third to half of the vegetation in any one year, and avoid the spring and fall waterfowl migrations and the summer nesting season. See *Conducting Prescribed Burns* in the suggested reading list for information on planning and conducting a safe prescribed burn.

## Additional Management Tips

Although, as discussed previously, planting or seeding in new plants may not be needed, sometimes you can improve existing aquatic habitat by introducing plants.

Plugs of wetland plants can be planted into the site. Seed may be planted or scattered, but most emergent plants need mudflats to sprout on, so don't scatter seed into the water. Plants and rootstocks may be planted right into the mud, even under water, provided the water isn't too deep for the plant.

Introducing plants or seeds after a site has dried up and been burned is a recommended method. The seedbed is exposed and easily planted once existing vegetation has been removed. Also, interseeding into a thick stand of cattails, cordgrass, or sedges will likely be unsuccessful, even if the above-ground portions of the plants have been temporarily eliminated through burning. The plants will return in the spring with vigor and quickly outcompete most plants you've tried to introduce. Plant only in areas with sparse vegetation.

Planting annual food plants such as millet or buckwheat to supplement natural moist-soil plants on mudflats is a common way to increase the abundance of food available for waterfowl in the fall. Be sure to concentrate on planting annuals, and do not plant non-native plants that are known to spread and cause problems to natural communities.

## Suggested Reading

*A Field Guide to the Wetlands of Illinois.* 1988. Illinois Department of Conservation.

*Conducting Prescribed Burns.* W. E. McClain. 2003. Illinois-Indiana Sea Grant College Program, University of Illinois.

*Illinois Landowner's Guide to Amphibian Conservation.* R. E. Szafoni, C. A. Phillips, S. R. Ballard, R. A. Brandon, and G. Kruse. 2002. Special Publication 22, Illinois Natural History Survey.

*Illinois Wetland Restoration and Creation Guide.* A. M. Admiraal, M. J. Morris, T. C. Brooks, J. W. Olson, and M. W. Miller. 1997. Special Publication 19, Illinois Natural History Survey.

*Management of Small Lakes and Ponds in Illinois.* 1995. Illinois Department of Natural Resources.

*Waterfowl Management Handbook.* 1988. Leaflet No. 13, U.S. Fish and Wildlife Service, Washington DC.

*Waterfowl of Illinois: Status and Management.* S. P. Havera. 1999. Special Publication 21, Illinois Natural History Survey.

FLOODED BOTTOMLAND FOREST

*Chapter 6*

# *Cropland* and Other Agricultural Areas

*Though Illinois has traditionally been known as the Prairie State, agricultural land has become our dominant habitat type. Cropland primarily occupies areas of Illinois that were formerly prairie. Today, 85% of Illinois land is farmland; of that, two-thirds is devoted to corn and soybeans. By comparison, remnants of the original prairie occupy less than one-tenth of 1% of our state.*

White-tailed deer
Previous page: left, ring-necked pheasant; right, coyote

Some think that croplands have little to offer wildlife. Indeed, the most intensively farmed areas have little but miles of row crops broken up only by an occasional pocket or corridor of woodland, grassland, or wetland. But some species do use croplands as a primary habitat. And, with just a little accommodation, cropland can provide habitat for many species. See Table 6.1 (p. 133) for selected species that frequent croplands. Providing more grassland, woodland, and wetland habitat may seem the logical choice for landowners interested in wildlife. But because cropland occupies more than three-quarters of our state, any realistic focus on managing for Illinois wildlife must include agricultural lands. Moreover, improving habitat on farmland will benefit wildlife in nearby grasslands, woodlands, and wetlands.

## What Defines Cropland Habitat

Croplands include fields whose primary use is producing row crops (primarily corn and soybeans), small grains (primarily oats and wheat), fruit and vegetable crops (strawberries, tomatoes, green beans, melons, etc.), hay crops, forage, and orchard crops. These fields are regularly disturbed by activities such as planting, tilling, harvesting, mowing, and grazing.

## Cropland Habitat Issues in Illinois

Historic environmental issues of cropland habitat differ from those related to other habitat types. Cropland is the only habitat that occupies more acreage in Illinois today than it did 150 years ago. In fact, outside of small agricultural fields and plots developed by Illinois'

Native Americans, cropland was non-existent 250 years ago. While loss of habitat acreage isn't a problem for wildlife using cropland like it is for species dependent on wetlands, woodlands, and grasslands, changes have occurred in cropping systems since the 1960s that have negatively affected wildlife. These changes are significant because they have occurred over a very large land base. Three changes are especially important:

1. The types of crops grown
2. The intensity of cropping
3. The management of agricultural lands

## *Types of Crops Grown*

From the early days of Illinois agriculture until the middle part of the 20th century, a broad mixture of crops was grown. Fields were frequently rotated, from hay, to corn, to small grains such as wheat and oats, and back to hay again. Both temporary and permanent pastures also formed part of the cropland mosaic. These early agricultural practices actually increased numbers of some species, such as prairie chickens, loggerhead shrikes, and Bobwhite quail.

This crop rotation and mixture of uses created habitat diversity on the landscape. The increased habitat variety in turn resulted in a diverse food chain base—insects, spiders and other invertebrates,

DIVERSE AGRICULTURAL LANDSCAPES OFFER WILDLIFE MORE OPPORTUNITIES.

American Goldfinches

rodents, reptiles, and amphibians—and ultimately provided for a variety of other wildlife, such as birds and larger mammals. The seasonal changes in the crops also provided cover that was disturbed at varying intervals, giving many wildlife species a better chance of nesting, raising young, and finding undisturbed habitat nearby when a particular site was tilled or harvested.

Today, two-thirds of our croplands grow only corn and soybeans. Neither crop can offer the diversity of wildlife cover provided by hay, pasture, and small grains. Hay and pasture, in particular, once provided a partial substitute for the millions of acres of native prairie that were lost. Corn and soybeans simply can't furnish the same type of habitat.

The hay-field crops grown in Illinois today also differ from those of the late 1800s and early to mid-1900s. Much of the hay produced now is alfalfa rather than the clover–grass mixtures of earlier days. Like corn and soybeans, alfalfa as a single plant provides for a less-diverse insect community. Crops grown as monocultures also provide less variety of post-harvest waste grain and seeds for wildlife consumption. In addition, many permanent pastures have been converted to monotypic tall fescue that is too dense for most wildlife to use.

## *Intensity of Cropping*

Though cultivated acreage increased steadily in Illinois after the invention of the steel moldboard plow in 1837, it has been only since the 1960s that the intensity and scope of agricultural land use—for example, "fencerow-to-fencerow" farming—have increased dramatically. This change has occurred as a result of several interrelated factors.

Machinery has gotten much larger, faster, and more efficient in recent decades. As a result, farmers can plant, harvest, and till more acres of row crops in less time. Another important factor has been federal agricultural policies, which have encouraged farmers to increase feed-grain output. As a result, more acres of pasture, hay, and small grains have been converted to row crops along with shrubby fencerows, odd grassy areas, and hedgerows, making field sizes much larger, with little or no edge or interspersed habitat.

These changes have reduced the overall amount and quality of habitat for wildlife that once thrived along the edges of crop fields or traveled between the croplands, grasslands, woodlands, and wetlands. Also, "wet spots," the ephemeral wetlands, continue to be drained in agricultural fields, further reducing the potential value of cropland to wildlife.

**Table 6.1 Selected Wildlife That Use Cropland Environments**

| Species | When they use | Why they use | What you can do |
|---|---|---|---|
| Red fox | Year-round | Search for small mammal prey | Plant and maintain grassed waterways and filter strips where prey can live. Protect adjacent areas where denning occurs. Leave in fence lines and farm lanes for travel routes. |
| Fox squirrel | Summer and winter; diurnal | Feed on standing corn in summer and waste corn and soybeans in winter | Leave in Osage orange hedgerows. Leave mature trees in fencerows for denning and nesting. Allow trees to grow in fencerows. |
| Ring-necked pheasant | Year-round | Find escape cover in summer, feed on waste grain in winter | Practice no-till as well as other forms of conservation tillage. Leave a few rows or small patches of grain unharvested. Plant and maintain grassed waterways and filter strips and leave unmowed, April 1 to August 1. |
| Horned lark | Year-round in all types of cropland, including crop stubble | Nest early spring before planting, and forage for insects, spiders, seeds; typically finish nesting by the time of spring planting | Reduce or eliminate fall tillage, especially plowing. Use no-till in spring. |
| Red-tailed hawk | Year-round, although the winter birds are usually different from summer birds | Hunt along crop field edges for small mammal prey, as well as in crop stubble and grassy areas | Maintain grassy areas along edges or in waterways which will support small mammal prey. Use filter strips that provide habitat for prey. Leave trees in fencerows for perching. Construct artificial perches. |
| American kestrel | Year-round | Hunt along crop field edges for small mammal prey, as well as in crop stubble and grassy areas | Maintain grassy areas along edges or in waterways which will support small mammal prey. Install nest boxes and protect boxes from predation. Plant and maintain filter strips that serve as prey habitat. |
| Garter snake | April through October; diurnal | Travel through, looking for insect prey | Practice no-till farming. Allow some weed growth and insect presence in field borders, grassed waterways, and odd areas next to crop fields. |
| Monarch butterfly | Summer; diurnal | Feed on milkweed, lay eggs | Allow some milkweed growth in field borders, grassed waterways, and odd areas next to fields. |

Soybeans no-tilled into corn stubble

The move from legume mixtures and legume–grass hay mixtures to alfalfa has increased disturbance of hay fields. Two or more cuttings of alfalfa can be taken each year. And alfalfa can be cut earlier in the year, coinciding with the peak nesting season for most grassland wildlife. This increased periodic disturbance has significantly decreased haylands' value to wildlife.

## *Management of Agricultural Lands*

The primary management practices that can influence cropland's value to wildlife are tillage, pesticide application, and mowing. Conservation tillage is now widely used on Illinois croplands. The residue from previous years' crops remains on the surface and provides significant habitat to many wildlife species throughout the year. And a food source is provided for many animals when the post-harvest waste grain is not turned under in the fall. Crop stubble also distributes snow more evenly in the winter, preventing heavy accumulation of snow in protective wildlife cover such as fencerows, grassed waterways, drainage-ditch banks, and terraces.

However, millions of acres of Illinois cropland continue to be tilled more extensively than is needed for optimum crop production. Besides eliminating valuable food and cover for wildlife, extensive tillage often causes off-site damage to wildlife habitat. It promotes soil erosion and causes sedimentation and pollution in aquatic habitats.

Integrated pest management strategies are being used over more acres, which may help reduce pesticide use over the long term. However, many pesticides are much more potent, and the use of pesticides still warrants concern. Rodenticides and insecticides have caused the most significant negative impacts to wildlife, and as a result they are now highly regulated. Many cases of poisoning have occurred in a variety of species, from eagles, songbirds, and game birds to various

Bumblebee

mammals, as an indirect result of pesticide application. Eggshell thinning in raptors, caused by the now-banned DDT, is a classic example.

As predators higher in the food chain consume dead or dying invertebrates or vertebrates targeted in a pesticide application, the pesticide may be passed on to the animal eating the poisoned bug or rodent. Immediate toxic effects may result, or there may be a delayed result from the cumulative effect of eating numerous poisoned prey.

Insecticide application also reduces the overall number of insects, reducing the food base for many wildlife species. A general insecticide application can also negatively impact crop fields themselves, because many insects that are beneficial (because they consume harmful insects or act as pollinators) are destroyed in the process. While some of the most persistent insecticides, like DDT, have been banned in the U.S., inappropriate use of approved products still impacts wildlife.

Herbicides used to control grasses or broadleaf weeds are not targeted to destroy insects, spiders, or vertebrate animals, but they may injure or eliminate some species. And eliminating all weeds destroys host plants for many beneficial insects. Destroying all weeds also removes potential food sources (seeds) and possible nesting sites for some wildlife species.

Nest and eggs of horned lark

Mowing in waterways, terraces, odd areas, field borders, roadsides, or crop fields reduces wildlife abundance, especially if done during nesting season. Keeping these areas visually tidy requires early and frequent mowing. Nesting cover, foraging cover, and often incubating adults, nests, and young themselves are destroyed when mowing is done during prime nesting and brooding season (April 1 to August 1). Conversely, late-season mowing (after September 15) destroys important winter cover and food for many species. Too much unnecessary mowing has hurt wildlife.

Horned lark

Some species have actually increased in the last 40 years in response to the growth in Illinois' row-crop acreage. Apparently they are less affected by cropping systems, pesticide use, mowing, and tillage practices. Among these are red-winged blackbirds, killdeers, and horned larks. But the great majority of Illinois native wildlife species have decreased as a result of agriculture's intensification and mechanization.

While most agricultural land in Illinois will never return to prairie, forest, or wetland, moderating some of the more recent changes in agricultural land management to address wildlife and other environmental considerations is essential to our wildlife's

ALWAYS CONSIDER HOW YOUR PROJECT WILL BENEFIT WILDLIFE IN THE SURROUNDING LANDSCAPE.

future. Even small changes, when adopted on large cropland acreages, can have a positive impact on the future of Illinois wildlife and can provide improved environmental health for humans.

## How You Can Help Cropland Wildlife

Illinois landowners can help wildlife use cropland by protecting and properly managing cropland and adjacent habitats. But before you embark on a project to do so, here are some things to consider.

### Management Considerations

When biologists create a management plan for a particular site, they think on two levels. The landscape perspective considers how a particular habitat site, or "patch," fits into the surrounding landscape. The patch perspective considers the management of habitat within the site itself. The following sections summarize these considerations, and chapter 2 provides further insight.

### Landscape-Level Management

No field or piece of land exists in a void, so it is important to consider the bigger picture of how a particular field or patch fits in with the surrounding landscape to provide regional habitat. The animals and plants in and around a site are affected by and interact with the surrounding landscape, and they don't recognize human-designated boundaries. In any cropland management effort, be sure to evaluate landscape-level considerations.

Regarding patch size, fragmentation is not an issue with cropland as it is with other habitats. The few wildlife species that use cropland for a primary habitat can prosper as well on a few acres as on several hundred. Generally, though, the smaller an agricultural field, the better it is for a spectrum of wildlife, especially if it is interwoven with other beneficial habitat.

Connectivity and adjacent habitats both come into play in improving cropland for wildlife uses. Depending on your management objective, croplands can be used to augment adjacent habitats, or other adjoining habitats can be developed to enhance a crop field. For example, studies have shown that grassy areas being managed for species that need large expanses of grassland provide more suitable habitat if they are adjacent to open cropland instead of forest. Likewise, adding a strip of woody cover to a crop-field border would benefit edge species. It is important to consider the proximity of a patch to other habitats and how to incorporate more interspersion of different habitats with the cropland.

## *Patch-Level Management*

Once you have evaluated large-scale considerations, you need to determine the management of the site or field itself. Two criteria must be addressed on any crop field: disturbance and plant-species diversity. The field will not provide suitable habitat for wildlife if the minimum standards are not met for one or both of these.

### Disturbance

Different from the other habitats discussed in this book, artificial disturbance is actually a regular process on agricultural lands; in fact, it's what defines this habitat type. So while preventing disturbance is not a goal, appropriate applications of disturbance—primarily timeliness and amount—need to be considered when managing for wildlife. It is important to follow the guidelines for disturbance timing, intervals, and amount in the subsequent section on agricultural land management.

BOBWHITE QUAIL

RED FOX PUPS AT PLAY

### Plant-Species Diversity

Croplands by nature do not contain a very diverse plant community. Increasing plant diversity in croplands will greatly benefit wildlife. This could mean planting companion crops, creating strips of alternate vegetation, or planting different crops in smaller management units. Rotating crops every year or two years will also improve plant-species diversity on a year-to-year basis.

## *Agricultural-Land Management for Wildlife*

The management of agricultural fields is addressed in three sections: croplands, haylands and pasture, and orchards. Minimum criteria for interspersed or adjacent cover, appropriate management of disturbances, and plant-species diversity need to be incorporated in any agricultural field for the site to provide suitable wildlife habitat. Options to provide enhanced wildlife value are also discussed.

POST-HARVEST WASTE GRAIN FROM CORNFIELDS IS AVAILABLE TO WILDLIFE IF FIELDS ARE NO-TILLED.

## *Cropland Management*

Fields that contain harvested grains or specialty crops should meet the following minimum standards: They should contain at least 2.5% (one acre per forty acres of cropland) interspersed or adjacent wetland, woody cover, or grassland that is managed for wildlife as discussed in the chapters on those habitat types. Fields need to retain a minimum of 30% plant residue until the next year's or next season's crop is planted. Rotate crops at least every three years. Although food is usually not the limiting factor in cropland settings, wildlife may be attracted to unharvested crops or food plots.

Wildlife value can be enhanced on croplands by increasing the amount of interspersed or adjacent cover, increasing the amount of residue, rotating row crops with small grains and grasses and legumes, and rotating crops more frequently. In addition, wildlife value increases with the reduction or elimination of pesticides. Always employ crop scouting to determine specific pest threats and the appropriate pesticide use (if any) to target only those pests that will produce significant economic impact.

Other specific activities to improve interspersion and increase residue are discussed in the later section on additional wildlife management (p. 141).

## *Hayland and Pasture Management*

Fields that contain harvested grasses and legumes or pastures that are intensively grazed should meet the following minimum standards: They should contain at least 2.5% (one acre per forty acres of pasture or hayland) interspersed or adjacent wetland, woody cover, or grassland that is undisturbed or managed for wildlife as discussed in the chapters on those habitat types. For hay fields, the preferred undisturbed cover is grassland.

Avoid mowing during late incubation (June 10 to July 1) if at all possible to avoid nesting hen pheasant losses. Also, a strip of uncut hay fifty feet wide should be left around all hay-field margins. This may be cut after August 1. For small hay fields, you can increase wildlife survivability by walking the field before cutting to flush out nesting birds and even marking nest locations so they can be avoided during cutting.

Intensively grazed pastures (those that do not meet the criteria in "Grassland Protection With Light Grazing" in chapter 3) should be set up on a rotational grazing system, allowing at least a quarter of the pasture to remain ungrazed for one month during the growing season. Stocking numbers and length of grazing period should result in no more than half of the total annual plant production being removed from the pasture, with vegetation heights never falling below six inches for cool-season pastures and eight inches for warm-season pastures.

WHITE-TAILED BUCK DEER IN VELVET

One way of producing a maximum forage crop while improving wildlife benefit is to raise and harvest native warm-season prairie grasses such as switchgrass, eastern gamagrass, and big bluestem. Studies have shown that nutritional values of these grasses are similar to traditional non-native grasses such as smooth brome and orchard-grass. And the same tonnage of hay can be obtained by just one annual cutting of these grasses as opposed to three or four cuttings of traditional hay like alfalfa, resulting in less disturbance to wildlife. The timing, too, benefits wildlife. The first cutting of prairie hay usually isn't done until early July, reducing disturbance to nesting wildlife. Remember that native warm-season grasses are cut higher (above the growing point).

KILLDEERS

Native warm-season grass pastures have advantages for both producers and wildlife. Because the native grasses do most of their growing in the warm months of late May through August and grow much taller than traditional cool-season grasses, they create good cover for wildlife during peak nesting season. Livestock can graze a cool-season pasture from spring through late June to early July and then be switched to the warm-season pasture. Remember that cool-season grasses have a "summer slump" in nutrition.

Assuming that the native warm-season prairie grasses haven't been burned, wildlife can nest and often fledge before livestock are allowed on the native warm-season pasture. And once the livestock are present, the tall grass gives the remaining active nests a better chance to survive. The rested cool-season pasture can then provide brood habitat for late or second-try nesters. Producers benefit because the use of forages at their peak production allows more livestock to graze a smaller area while still receiving adequate nutritional value. Studies show that rotational grazing using both cool- and warm-season grasses produces higher annual weight gain per acre than cool-season pasture alone, thus reducing the costs of production.

See chapter 3 for more information on establishing native warm-season grasses. Seeding rates given in Table 3.3 (p. 60) are for wildlife plantings. Forage plantings require a higher rate. Additional specific activities to improve interspersion of other habitat in pastures and haylands are discussed in the upcoming management section.

COOL-SEASON PASTURE GRASSES SHOULD NOT BE GRAZED BELOW SIX INCHES.

## *Orchard Management*

Fields that contain woody fruit crops should meet the following minimum standards: They should contain at least 2.5% (one acre per forty acres of orchard) interspersed or adjacent wetland, woody cover, or grassland that is undisturbed or managed for wildlife as discussed in the chapters on those habitat types. The vegetation between trees or shrubs should remain unmowed after fruit harvest to provide winter cover. Pesticide use should be kept at a minimum. Crop scouting should be done to determine any problem pests, and pesticides should be used only on those pests that will have economic impact.

Wildlife value in orchards can be enhanced by leaving vegetation between tree or shrub rows undisturbed during the nesting season (April 1 through August 1). Consider using companion plantings between rows to repel or deter known invertebrate pests and thus provide cover and reduce pesticide use.

More details on management activities to improve interspersion are provided in the next section.

## *Additional Wildlife Management on Agricultural Land*

Various activities can create additional wildlife cover on farmland. Many have the added benefit of protecting wildlife habitat away from the agricultural site, primarily by reducing or eliminating soil and pesticide movement into aquatic habitats. However, many of these practices, if not managed properly, could function to protect off-site habitat but not provide any habitat on the agricultural field itself. For instance, a legume–grass buffer strip forty feet wide can protect a nearby wetland from sedimentation and pollution, but if it is mowed several times during the nesting season, it would provide little benefit to wildlife living near the buffer strip.

The following discussion mentions off-site benefits, but it also focuses on how to manage the land for on-site wildlife benefit. Contour strip cropping, terraces, grassed waterways, and tillage and residue management apply primarily to row-crop, small-grain and specialty-crop fields, though they can be used on other agricultural lands. Field borders, filter strips, food plots, and miscellaneous applications apply not only to cropland but also to pastures, haylands, and orchards.

EASTERN MEADOWLARK

## *Contour Buffer Strips and Terraces*

Contour buffer strips and terraces can be used on sloping cropland to decrease or eliminate soil erosion. The tops of unfarmable terraces and the alternate strips between row-cropped strips are planted to some type of permanent herbaceous vegetation, usually grasses or legume–grass mixtures. The size of the strip, what is planted, and how it is managed determine its value to wildlife. Strips should be at least ten feet wide, but wider is better. Strip width and number depend on slope and must be compatible with machinery width.

Types of grasses that can be planted are discussed in chapter 3. Terraces are seldom mowed and are good sites for establishing native warm-season grasses and forbs. To provide adequate wildlife habitat, the strips should not be mowed during the nesting season (April 1 to August 1). Also, vegetation should be left unmowed after September 15 to provide winter cover. Terraces and contour buffer strips should be designed by an NRCS soil conservation professional.

Ring-necked Pheasant

## *Grassed Waterways*

The primary purpose for constructing a grassed waterway is to prevent gully erosion by providing a stable, vegetated channel for water to exit a crop field. If shaped and grassed to provide for infrequent mowing, a waterway can also provide some cover for wildlife. Grassed waterways are often less valuable than other grassy cropland applications because periodic inundation prevents ground-nesting species from using the areas successfully. Some species can nest on the upper edges of wide waterways, and certainly many animals can use them for foraging, concealment from predators, and travel lanes.

Fescue and reed canary grass are widely used in waterways, but they are the poorest cover choices for wildlife. Both grasses get extremely thick and are difficult for many species to travel through. Other grass mixtures of species such as brome, redtop, orchardgrass, and the moisture-tolerant native prairie grasses like switchgrass, Indian grass, and big bluestem are much more valuable to wildlife. Follow the guidelines in chapter 3 on seeding recommendations. If a

From his years growing up on a farm, Jerry Heinz remembers livestock, a variety of grain and legume fields, fencerows, and lots of wildlife. He always "just assumed" the wildlife would be there, but over the years he witnessed the disappearance of small fields, fencerows, and, inevitably, wildlife. Jerry still lives on the Champaign County farm where he grew up in east-central Illinois, and fifteen years ago he decided there had to be a better way to farm. Today he has 900 acres in crop production and 50 acres in conservation programs. He also has a smaller farm operation in far southern Illinois (Union County) with about 95 acres in the Conservation Reserve Program (CRP) and 180 acres in the Wetlands Reserve Program.

Jerry signed up for his first CRP acres in 1989. He planted 6-1/2 acres in a switchgrass filter strip, and by 1990 he had his first Acres for Wildlife sign, something he still treasures. Currently he has enrolled all the ground that is eligible in the CRP. He has filter strips, a shallow water wetland, waterways, windbreaks, shelterbelts, and trees. He also has annual food plot acreage, which he plants on his own because it is not part of any program.

Jerry's philosophy has been simple: "It's a good thing to do. The owner knows the land the best and needs to go with what's right. You know what you want and you should go with that." He advises landowners to "use the different agencies and learn how to use the programs. That way you can do what is best for your farm."

While Jerry has seen an economic benefit from the CRP, more important to him is the conviction that he has brought back both the quality and diversity of wildlife. He has seen an explosion of wildlife on his land. Pheasant sightings have gone from a few birds to as many as fifty in a year. Badgers have found a home here, while the groundhogs, which were a problem, have disappeared. Owls are nesting in an old silo, and rabbit and hawk numbers have skyrocketed. One of his favorite sights in the fall is watching Canada geese and mallards drop straight into the corn stubble. "I wasn't a 'waterfowler' growing up, but I'm hooked now!"

When asked "What do the neighbors think?" Jerry smiles before responding. Some neighbors just shake their heads, but others have stopped to ask questions. On his Union County property, when water grassed waterway receives repeated traffic from farm machinery, especially early in the growing season, nesting success will be low. Consider mowing such high-traffic areas early, and keep them short to discourage nesting attempts. Cease mowing once the last spray or tillage trip is made. If a grassed waterway does not receive repeated farm traffic, leave it unmowed during the nesting season (April 1 to August 1) and after September 15.

Grassed waterways should be designed by an NRCS soil conservation professional. Cost-sharing may be available. If you plan to leave your waterway unmowed much of the year, make that known before the design is created. Unmowed grass slows water flow more than mowed grass, a factor that will need to be considered.

### *Field Borders and Filter Strips*

Strips of cover along the edges or borders of a crop field can provide significant wildlife habitat while protecting adjacent or nearby watercourses from sedimentation and pollution. At a minimum, field borders and filter strips should be at least ten feet wide; however, for wildlife use, wider is better. The NRCS has standards and specifications for both the widths and vegetation selections for

control structures were being installed for wetlands, the former landowner wondered "what these kids were up to wasting prime farmland." But this past fall that same skeptic came to watch the ducks.

Perhaps the best indicator of what Heinz's neighbors think is the choice that some have made to duplicate his conservation practices on their own land. As a result, the water quality in the area of his Champaign County farm has improved.

Jerry is modest about his accomplishments. "I can't take credit for how much clearer the water runs or how many more wildlife species call this place home. The credit goes to all the farmers in the watershed who are doing the same, who have made conservation a priority." Yet Jerry Heinz is leading by example—creating habitat where none existed and helping wildlife where he can—and he's doing it the old-fashioned way, one farm and one farmer at a time.

*Susan L. Post*

*Jerry Heinz*

filter strips. Often planting a field border or filter strip requires taking only a small portion of cropland out of production to widen a drainage ditch border, a stream bank, or the perimeter of a field. The wildlife benefits will far outweigh the small reduction in crop acreage. Any of the mixes listed in chapter 3 can be used for field borders and filter strips.

Again, the key to wildlife value is properly managing disturbance to the cover after establishment. Mowing or burning should not be done during the nesting season (April 1 to August 1). Avoid using field borders as travel or access lanes during the nesting season. Leave tall vegetation standing over winter for cover. Burning, mowing, and light discing outside of the nesting season can all be done to prevent woody encroachment.

In some cases, woody vegetation may be desired in a field border or filter strip. It is recommended that a light seeding of grasses and legumes be planted first. Shrubs or trees should then be planted into the herbaceous vegetation. Low-growing native shrubs (such as dogwood, hazelnut, elderberry, American cranberry, and black chokeberry) provide substantial wildlife cover and erosion protection but have minimal moisture and shading impacts on adjacent cropland. If taller deciduous or evergreen trees are desired, tractor-pulled root plows may be used to keep tree roots from extending into the crop field and taking moisture and nutrients from row crops, small grains, or forage crops. Even though field windbreaks may sap moisture and nutrients from adjacent crop rows, they increase total yield over the entire field. These windbreaks reduce wind stress, particularly from dry summer winds.

White-tailed buck deer

Chapter 4 provides recommendations for tree and shrub plantings. If you currently have a woody fencerow or field border, consider retaining it and using a root plow to cut any competing roots. This will allow the continued coexistence of valuable wildlife habitat and the farm's commodity crops.

## Roadsides

Roadsides that are managed for wildlife provide nesting habitat for pheasants and other grassland birds. Cottontail rabbits and voles increase on these roadsides, and in turn attract red-tailed hawks and American kestrels. While roadsides occupy only 1% of the Illinois countryside, when properly managed, they can provide significant wildlife habitat.

Seeded roadsides provide much better habitat than do unseeded roadsides, which are often mostly bluegrass, broadleaf weeds, and fescue. Roadsides can be tilled and planted with mixtures of smooth brome, alfalfa, red clover, timothy, orchardgrass, redtop, and lespedeza. Native warm-season grasses can be used if special attention is given. Species such as big bluestem and Indian grass are tall and can reduce traffic visibility. Tall vegetation left standing over winter may also lead to snow-drifting problems. However, these concerns can be remedied. Native warm-season grasses should not be planted near intersections or farm-lane entrances where traffic safety is a concern. Areas prone to snow drifting can be mowed after the grasses enter dormancy in late fall.

Native cool-season grasses afford some roadside advantages not offered by their warm-season counterparts. Virginia and Canada wild rye are shorter than most warm-season natives and they can be mowed earlier, which addresses visibility and snow-drifting concerns. In addition, they grow earlier in spring than warm-season species, thus providing earlier cover and nesting habitat.

As with all grassy areas managed for wildlife, roadsides should not be mowed until after August 1. *Delaying mowing until after the nesting season is the single most important thing farmers and road maintenance personnel can do to benefit grassland wildlife.* Delayed mowing not only helps wildlife, it saves time and fuel. However, if roadside vegetation presents a visibility problem, such as at an intersection, the area should be mowed early and often during the growing season. As well as improving visibility, early and continuous mowing discourages nesting activities that would almost certainly end in failure.

Landowners can receive assistance in managing roadsides by contacting district wildlife biologists or the Roadsides for Wildlife program offered by IDNR. Landowners should make every effort to improve roadside habitat because these areas will remain an important habitat component, especially for grassland birds.

AMERICAN KESTREL

## *Tillage and Residue Management*

Although no-till has become widely used in Illinois, there are still more acres where reduced tillage should be adopted. Just a small change in tillage operations—leaving more residue on a crop field—provides significant wildlife and soil conservation benefits.

PROPERLY MANAGED ROADSIDES OFFER WILDLIFE A HABITAT OASIS IN AREAS THAT ARE FARMED INTENSIVELY.

Woody cover at a field edge offers winter protection.

Untilled stalks of crops such as corn and sorghum provide some winter cover and help distribute snow evenly over the field. Leaving winter wheat stubble untilled in summer provides brood and roosting habitat for some re-nesting and late-nesting birds. No-till provides even more benefits since soil microorganisms and invertebrates such as earthworms are allowed to increase, providing a food source for many animals. No-till also leaves any post-harvest waste grain on the surface to be used as winter wildlife food. While a minimum of 30% residue must be left on the field at all times for adequate habitat, no-till or reduced-till that leaves larger amounts of residue will provide enhanced value for wildlife.

Another way to control erosion, decrease certain weeds, increase crop-field nutrients, and provide significant habitat for wildlife is to plant companion crops and winter cover crops. Winter cover crops are usually legumes, such as hairy vetch, that grow quickly. In many instances, the spring crop can be no-tilled right into the cover crop. Companion crops are planted between the rows of the commodity crop, and their coverage keeps weeds from establishing or flourishing. Both companion and winter-cover crops provide wildlife benefit—primarily increasing foraging and nesting opportunities—by reducing disturbance (tillage and pesticide application) and increasing plant-species diversity.

## *Food and Cover Plots*

One simple way to provide additional wildlife habitat in a crop field is to leave a portion of the planted crop unharvested through the winter for food and cover. Or you can plant grains and legumes specifically for a food and cover plot, providing both valuable shelter from inclement winter weather and an emergency or supplemental winter food source.

Food and cover plots can be planted in a variety of locations with various plant combinations. Here are some basic design tips:

- The predominant winter winds in Illinois originate from the north and west. When possible, place food and cover plots, especially smaller plots, on the south and east side of other protective cover, such as grassland or woody habitat.
- If your property is in a flat, intensively farmed portion of Illinois where there may not be an opportunity for protection from winter winds, plant a large block rather than a strip of food and cover plot. Wildlife can find some shelter in the interior portions of the plot when needed. Blocks need to be at least 300 feet by 300 feet to provide effective winter shelter; larger is preferable.
- Plant a variety of grains and legumes in the plot. Corn and grain sorghum usually stand erect in snow, providing good cover. Soybeans provide a ground-level food source, whereas corn provides an above-the-snow emergency food source. See Tables 6.2 and 6.3 for specific recommendations.

MOURNING DOVE

RED FOX

- Plant or drill food-plot rows relatively close together (no more than fifteen inches apart) to provide more compact cover. Plant different crops in separate rows and blocks and consider weed-control options in the design. Weed control is important only as plantings are getting established. Weedy plants can be a desirable component once food and cover plots are established.
- Food and cover plots may be the best wildlife practice for annual set-asides sometimes offered in farm programs.

## *Miscellaneous Applications*

Sometimes odd areas are created by an asymmetrical field or machinery turn-around points. These areas represent yet another opportunity to create valuable wildlife habitat. Odd areas can be planted with grass–legume mixtures, shrubs, and trees or as food plots, although food plots may not be needed near grain fields with

**Table 6.2 Food Plot Plant Selection and Management Guide**

| Species | Planting dates | Seeding depth | Seeding rates | Weed control | Comments |
|---|---|---|---|---|---|
| Corn | Late April–early May | 1.5 to 2 in. | For 30-in. rows: 1 plant per 8 to 10 in. (20,000–25,000 plants/acre) | Consult locally for herbicide recommendations; cultivate twice if no herbicide is used | Best all-around food for wildlife |
| Milo (grain sorghum) | Late May–mid June (north) Mid May–late June (south) | 0.75 to 1 in. | For 30-in. rows: 4 to 6 plants per ft of row or drill at 8 to 10 lb/acre | Consult locally for herbicide recommendations; cultivate twice if no herbicide is used | Three to 4 ft tall; After corn, the best choice for winter wildlife when snow is on the ground; usable to more species if mowed or driven down, especially early in winter |
| Forage sorghum | Late May–mid June (north) Mid May–late June (south) | 0.75 to 1 in. | For 30-in. rows: 4 to 6 plants per ft of row or drill at 8 to 10 lb/acre | Consult locally for herbicide recommendations; cultivate twice if no herbicide is used | Same as milo but 5 to 8 ft tall; good for windbreak value |
| Sunflowers | Early April–early May | 1.5 to 2 in. | For 30-in. rows: 1 plant per 6 to 8 in. or drilled at 5 to 6 lb/acre | 2 pt/acre Treflan; preplant and incorporated; cultivate twice | For fall food only, use any commercial variety; preferred by mourning doves |
| Food plot mixes* | May–June | 0.75 to 1.5 in. | Broadcast at 20 lb/acre; drill at 10 to 12 lb/acre | No herbicide can be used; can rotary hoe early | Expect a weedier stand with mixed food plots; best where no post-plant management will occur |
| Wheat | October | 0.5 to 1 in. | Drill at 90 lb/acre | No herbicide necessary | For fall–winter browse and summer seed (doves) |

*Contact IDNR or other agencies for their food plot mixes.

*Seeding rates:* Rates are given for 100% germination. Adjust seeding rates based on germination of seed you are using.
*Planting method:* Test soil, fertilize, and prepare a good seedbed as for regular crops. Use a planter or small-grain drill with press wheels for larger-seeded grains. If broadcasting seed, till lightly to cover seed (and prevent wildlife from eating pesticide–treated seeds), then roll to firm seedbed. Control weeds with cultivation or herbicides as needed. Seed at proper rates. Overseeding causes stunting and poor yield.
*Fertilization:* Fertilize food plots as required based on soils. A good guideline if soil tests are unavailable is 20 to 50 lb/acre each of nitrogen, phosphorus, and potassium. Increase nitrogen to 50 lb/acre for sunflowers.
*Rotations:* Rotate plantings to prevent disease and increase stand vigor.

**Table 6.3 Food and Cover Plot Seeding Rates and Rotations**

| | Species | Pounds per acre |
|---|---|---|
| Grains | Milo | 8.0 |
| | Corn | 10.0 |
| | Sunflower | 7.0 |
| | Millet | 20.0 |
| | Cowpeas | 20.0 |
| | Soybeans | 30.0 |
| | Buckwheat | 50.0 |
| | Oats | 50.0 |
| | Wheat | 40.0 |
| Grasses | Bluegrass | 1.0 |
| | Timothy | 2.0 |
| | Redtop | 0.5 |
| | Orchardgrass | 3.5 |
| Legumes | Dutch white clover | 10.0 |
| | Ladino clover | 3.0 |
| | Alsike clover | 3.5 |
| | Red clover | 8.5 |
| | Alfalfa | 12.0 |
| | Birdsfoot trefoil | 6.5 |

Select from list. If using more than one type of grain, plant in separate areas of the unit. Plant approximately half the field to grains and half to legumes. Or, if possible, use a rotation of various grain and forage plants. Use the same number of plots as years in the rotation. See sample rotations below for suggestions.

| Rotation | Year 1 | Year 2 | Year 3 | Year 4 | Year 5 |
|---|---|---|---|---|---|
| 3-year | Corn and milo | Oats/legumes | Idle | | |
| 3-year | Corn and milo | Sunflower | Wheat | | |
| 4-year | Corn and milo | Oats/grass/legumes | Idle | Idle | |
| 5-year | Corn and milo | Sunflower | Wheat | Legumes | Idle |

Plant in spring, except for wheat. Plant sunflower in spring and wheat in fall of same season. Sunflower seed is usually eaten by migrants and gone by fall. Use oat nurse crop for first-year legumes and/or grasses.

MILO PLANTED IN A FOOD PLOT

considerable residue. Manage the area for nesting and wintering wildlife as discussed in the previous sections.

Another simple way to provide some habitat and help reduce pest problems at the same time is to install perch sites in or adjacent to a crop field. Predators of rodents such as hawks and owls use these perches when searching for prey. Aerial insectivores such as swallows need places to rest near their foraging area.

Many of the row-crop areas, especially in central Illinois, have few trees that can be used by these species as perches. Red-tailed hawks often sit on short fenceposts along highways because there are no other perch sites. Adding a couple of perches will provide additional habitat for these species and help reduce rodent and insect pests. See chapter 7 for more details.

## Suggested Reading

*Aldo Leopold: For the Health of the Land.* Edited by J. B. Callicott and E. T. Freyfogle. 1999. Island Press, Washington, DC.

*Field Office Technical Guide.* Natural Resources Conservation Service, Washington, DC.

*Illinois Agronomy Handbook (23rd edition).* 2003. Circular 1383, College of Agricultural, Consumer and Environmental Sciences, University of Illinois.

*Techniques for Wildlife Habitat Management of Uplands.* N. F. Payne and F. C. Bryant. 1994. McGraw-Hill, New York, NY.

Red-tailed hawk

*Chapter 7*

# *Special Features* and Supplemental Practices

*Reestablishing a mosaic of wildlife habitats takes time, effort, and a good amount of patience. Within limits, plants grow at their own predetermined rates, and we can do only so much to help them along. Plant communities, the basis of all habitats, develop slowly, and thus require patient effort. However, some special features—human-constructed parts of wildlife habitats—can be quickly established and give early results.*

EASTERN BLUEBIRD WITH INSECT ON NEST BOX
PREVIOUS PAGE: LEFT, BARN OWLS ON NEST BOX; RIGHT, CAROLINA WREN NEST IN FARM-SHED BUCKET

These "special features," as they are called, make up some of the structural components of habitats. Wildlife biologists have learned over time that some of these components can be easily built by humans, and that some wildlife species readily use these structures for nesting space or shelter—as long as they somewhat resemble their natural counterparts. The human-created structures most commonly used by wildlife include brushpiles, rockpiles, nest boxes, nest platforms and islands, perches, old buildings, and bridges.

In addition to producing relatively fast results, many of these special features can be incorporated into almost any habitat management plan. You don't need a particular soil type or a certain property size to install a nest box or build a brushpile. These features can be used in conjunction with nearly any land or wetland project.

It is important to understand, though, that while these special features may seem to offer a "quick fix" for wildlife, they should never take the place of good habitat management. They are to be viewed only as supplementary to your other activities. Nest boxes, brushpiles, and other special features can meet specific needs for certain wildlife, but to be successful these species need the other elements necessary for survival. For example, you could install scores

GARTER SNAKE

of wood-duck boxes, but if there is no appropriate wetland habitat nearby for brood rearing and foraging, wood ducks won't use the boxes. Likewise, dozens of brushpiles won't ensure a healthy rabbit population if there is no suitable nest cover within a quarter mile or so.

As long as special features are used in appropriate contexts, they can provide great satisfaction to landowners anxious to see some tangible results for their efforts. Not every special feature would be appropriate or useful for every property, but all landowners should consider adding nest boxes and brushpiles, which are almost universally appropriate and beneficial.

## *Brushpiles*

Brushpiles are the easiest, cheapest, and quickest way to create "instant" cover. They can usually be constructed without any cost, they can be built quickly, and animals can begin using them as soon as construction is finished.

Nature creates her own brushpiles in forests and woody fencerows. When a tree topples, various sizes of limbs and other debris are usually piled randomly in and around it. Tree seedlings, shrubs, and vines spring up around the fallen tree. A natural brushpile is often where you find rabbit and Bobwhite quail hideouts, groundhog burrows, Carolina wren nests, and various reptile residences. Replicating this microhabitat is fairly simple.

### *Location*

Though brushpiles are beneficial in nearly any location, they are particularly valuable at the edge of adjoining habitats, such as between a woodland and a grassland. In a natural setting these areas are normally brushy, so placing a pile there imitates natural conditions. Brushpiles

COTTONTAIL RABBIT

Red fox

can also be valuable for escape cover in open habitats with short or sparse vegetation, such as cool-season grasslands and croplands. The piles can also provide critical winter cover in these areas, where little might exist otherwise. A brushpile near a food plot or water provides many animals the opportunity to feed or drink while having escape cover nearby.

There are some places you should not place a brushpile:

- Waterways, ditchbanks, ravines, and floodplains subject to frequent flooding. You do not want to negate the benefits to wildlife by flooding occupants out just when they may need cover the most.
- Areas that will include prescribed burning in their management regime. For the safety of both wildlife and the brushpile itself, you don't want the pile to be fuel for your prescribed burn. Most brushpiles would go up in flames immediately, especially if the wood has aged. If you are burning a woodland with only a thin ground-cover layer that won't produce a vigorous fire, it may be feasible to rake a firebreak around it and wet down the brushpile before burning to protect it. Still, if possible, try to route the burn away from any existing brushpile.

### *Size*

It can't be emphasized enough that brushpiles must be fairly large to provide effective cover. Animals seeking cover from enemies or inclement weather must be able to get far enough in and under the pile to have adequate shelter. A diameter of fifteen to twenty feet is considered a minimum size, and larger is even better. Materials should be piled at least six or seven feet high. It is much better to have one or two large piles than four or five small ones.

Don't be surprised when your brushpile "sinks" after four or five months. Brushpiles will continuously settle, and the bottom will decompose over time. Periodically add new material to the pile and it will continue to provide good cover indefinitely.

## *Construction*

As with a house built for human inhabitants, the quality of initial construction will determine a brushpile's longevity and utility. Start by building a well-spaced and sturdy base. Criss-cross the largest limbs available on the bottom, four to eight inches apart. This will provide tunnel spaces and easy access for small animals. Logs from rot-resistant trees, such as Osage orange, black locust, white oak, and eastern redcedar, make durable bases. Other usable materials for the foundation of a brushpile include fenceposts not treated with creosote, field tile, large stones, and existing stumps.

PRAIRIE VOLE

Once you've built the base, begin covering it with other branches, progressively smaller. This tight weave of brush above the base will form a sort of "igloo" of protection from rain, snow, and wind. It will also provide smaller prey species such as chipmunks, rabbits, Bobwhite quail, and lizards a safe haven from many predators.

BOBWHITE QUAIL

Materials for brushpiles are everywhere. Use the fallen limbs in your yard, windbreak, or woods. If you cut firewood or timber or are doing a selective thinning, use the brushy by-products for brushpiles. And leftover Christmas trees abound in January.

EASTERN FENCE LIZARD

## *Rockpiles*

Rockpiles serve a purpose similar to brushpiles, especially for smaller species such as snakes, lizards, and mice. Design should be similar to brushpiles, with larger rocks on the bottom to create a network of tunnels or spaces. The larger the rockpile, the better—two to three feet in diameter is a minimum size.

A rockpile can be placed anywhere, including in an area that will be subject to prescribed burning. But, as with brushpiles, route the burn away from the rockpile to minimize smoke and heat problems for current dwellers. Rockpiles are particularly valuable in woods for forest-floor dwellers. And unlike brushpiles, rockpiles can be used in ravines and other watercourses for animals like salamanders. Individual or piled rocks can also be used along the edges of marshes and ponds to provide hiding places for many aquatic species, including frogs and water snakes. When placed in the water with some surface exposed, rocks can provide basking areas for turtles and dragonflies.

The best source of material for your rockpiles is along roadway rock cuts, strip-mined fields, and quarries, where removing rocks is not disturbing an intact ecosystem. Rocks should not be taken from glades and rocky bluffs where they are already providing established habitat for a variety of species. These rocks are usually scattered and make up an important part of the local ecosystem. Rocks may be collected from talus slopes and rocky creeks only if taken in small quantities. Be sure to get permission from the landowner.

A VARIETY OF WILDLIFE USE NATURALLY OCCURRING ROCKY HABITAT.

## *Perches*

Artificial perches can be beneficial in areas lacking tall trees, towers, and power-line poles. Many hawks and owls, or raptors, need high perches where they can scan the landscape for small prey. Many older, tall trees have been eliminated in rural areas, forcing raptors onto short perches like fences, fenceposts, and young trees. If even these less-than-desirable perches are present, raptors won't stay in the area regularly.

Artificial perches can benefit both the raptors and the farmer. Several studies have shown significant reductions in rodent populations in the vicinity of recently erected artificial perches, such as in hay fields. Another excellent location is reclaimed strip mines.

At least nine species of hawks and owls that occur in Illinois use artificial perches. And though raptors may only occasionally be seen using a perch, it can be enjoyable to view the many other bird species that use it.

RED-TAILED HAWK

## *Location*

Perches should be placed in or adjacent to grasslands, crop fields, or wetlands. Do not locate a perch near any object that could entangle birds. Locate constructed perches at least 300 feet from each other and from existing natural perches.

## *Size and Construction*

Perches should be at least ten to twelve feet high, although twenty to thirty feet has shown better results for some raptor species. The platform or perch itself should be at least two inches wide by twelve to eighteen inches long, and at least two inches thick to provide support for larger raptors. Build perches with durable materials; either metal or wood may be used for the pole, but use wood for the perch itself.

Another way to make a perch is to create a snag and leave it standing at the edge of a crop field or grassland. But killing a live tree to create a perch is recommended only if the tree was slated for removal or creates a nuisance for farming operations. Live trees will be used by raptors as perches.

Providing perches for aerial foragers such as swallows and martins can also benefit both the birds and the farmer. Each swallow or

martin consumes countless insects every year, providing some relief from insect pests. Swallows and martins like to perch together and can often be seen sitting in large groups on telephone or electric wires. Similar perches can be constructed by running cable, rope, dowel rod, or other suitable material at least twelve feet above the ground between two trees or two posts.

These perches can be placed in or near a crop field, but they are even better located within a quarter-mile of a pond, lake, marsh, pasture, hay field, or undisturbed grassy area.

American kestrel with prey

## *Nest Boxes and Artificial Cavities*

Of the dozens of Illinois bird and mammal species that use cavities or tree hollows at some point in their lives, at least twenty-seven will use artificial nest boxes to raise young, roost, or seek winter cover. With cavity-rich old-growth forests on the decline in recent decades, and with farmers replacing wooden fence-posts with metal ones or eliminating them altogether, many cavity-nesting species are finding a shortage of accommodations. Several species readily use artificial structures, and some, such as the bluebird, have made a come-back largely due to the broad effort by Illinoisans to provide nest boxes.

While artificial boxes may be used permanently to attract species for

American kestrel nest box with metal guard on pole to deter climbing predators
Inset: American kestrel female incubating in nest box

viewing, set a long-term goal of providing enough large trees on your property to supply plenty of natural cavities. Artificial boxes don't always provide the same level of protection from competition and predators that natural cavities do. Also, the more nesting sites available, the better. And lack of natural nest cavities usually indicates lack of other habitat elements.

## *Design, Placement, and Maintenance of Nest Boxes*

Using a proven design for the desired species is essential. If you do not use an appropriate box size or the right diameter entrance for the hole, the desired species may not be able to use the box or may be subjected to competition from opportunistic or more aggressive species. An Illinois Department of Natural Resources (IDNR) publication, *Wood Projects for Illinois Wildlife,* has proven nest box designs for several birds and mammals. Many nature stores and nonprofit conservation organizations also sell boxes for individual species.

Once you've constructed or bought an appropriate nest box, it must be correctly installed and maintained. The box must be located both in the right type of habitat and at the right height to attract the target species. And seasonal cleaning is required of most boxes. Many species simply won't use a box with old nest material in it. The wood projects publication provides details on placing and maintaining nest boxes.

Also take measures to keep predators such as certain snakes and mammals from gaining access to the box. Placing metal guards around wooden posts or poles usually deters most climbing predators (see photo on previous page). Installing sections of PVC pipe (five feet or longer) over steel posts and occasionally greasing the pipe will also discourage climbing predators such as raccoons.

## *Artificial Nest Cavities*

A nest cavity can be created by killing an existing tree or inducing a live tree to form a hollow. If you have plenty of good-sized, healthy trees in a forest or fencerow and can afford to spare a couple, creating den or nesting trees can boost the habitat's value to wildlife. If there are no dead trees on or near your property, consider completely killing one or two live trees. A large, dead tree may actually contain several hollows, creating a wildlife "apartment." Girdling is an effective way to kill a tree while leaving it standing.

To create singular hollows in an otherwise live tree, try this: pick out a tree that is prone to forming cavities fairly quickly, such as a sycamore, silver maple, elm, or cottonwood. Choose one of the tree's larger branches at least ten feet above the ground. Cut the branch off about six inches from the trunk, creating a stub. To speed the cavity formation, chip or drill out some of the inside of the stub. The stub will prevent proper wound healing, eventually rotting into the tree. Be patient; it may take a several years to create a usable cavity.

FOX SQUIRREL AT DEN SITE

CANADA GOOSE WITH NEWLY HATCHED GOSLINGS

## *Nest Islands and Platforms*

Islands in ponds and marshes may be used by numerous species for nesting and foraging. A properly located island is especially beneficial to nesting waterfowl, such as Canada geese and mallards, because it provides some protection from predation. An island covered with and surrounded by vegetation also gives other creatures, such as frogs, additional hiding areas. And once constructed, an island is usually permanent and fairly maintenance-free.

If you are planning to build a pond or marsh, one or more islands can be easily integrated into your plans. Here are some guidelines:

- Generally, don't build more than two or three islands per acre of water. A smaller number is fine.
- Islands should be at least 100 square feet and rise at least three feet above the maximum high-water level to prevent flooding. Larger areas are better.

- At least thirty feet is needed between the island and shore; more is better. Between these points the water should be at least three feet deep. This may deter predators from swimming to the island to prey on nests and young.
- Make the island an irregular shape. Also, when creating an island in a pond, consider building part of it with gently sloping sides to create shallow water areas. This will produce aquatic vegetation and create a small marsh habitat around the island.

In an existing wetland, there is really no way to construct an earthen island without draining the wetland. But there are other ways to create nesting habitat for geese and other waterfowl. One of the easiest is to drop a large round hay bale on top of a frozen wetland in the winter. When the ice melts, the bale will sink. The trick, of course, is to place the bale where the water is shallow enough that a portion of the bale will stick out of the water once it has sunk.

A variety of designs exist for constructing wooden platforms for nesting waterfowl. Some of these platforms float; others are anchored in the mud with poles. The IDNR has designs available upon request.

Another way to construct a fairly permanent nesting structure for waterfowl is to use a large corrugated culvert or similar tile at least four feet in diameter. Submerge the culvert vertically into the wetland bed so that the top is at least three feet above the high-water level. Fill it with rock and then dirt, and plant a little grass and some forbs on top to provide nesting cover. This structure will offer protection from many predators, except for raptors and those animals adept at climbing out of the water and up a corrugated structure, like raccoons.

Virginia rail

## *Old Buildings and Other Structures*

A wide variety of wildlife use human-made structures, and certain species have flourished as a result of our developments. Some species, such as the state-endangered barn owl and the uncommon black vulture found in southern Illinois, have come to depend on our structures to replace lost habitat. Farm buildings, chimneys, abandoned houses, silos, road culverts, fenceposts, post piles, and bridges are used regularly by a wide array of wildlife.

Many of these structures in rural Illinois are symbols of days gone by, when numerous small farms dotted the landscape. Building materials were typically simple—primarily wood, stone, brick, and concrete. Since these materials were similar to those found in the nearby forest or woodlot, they created opportunities for nest and roost sites.

Many modern structures use materials that are new to wildlife. Today's farm buildings are usually made out of metal; many homes and sheds are covered with vinyl siding; fenceposts are metal rather than wood. With the advent of new building materials and designs, some of the wildlife that had adapted to our human-made world are having to learn new ways of making do.

If you have an old building on your property and no particular need for the space, why not leave it stand if it is not a public safety hazard? The same goes for wooden fenceposts and abandoned bridges. If a structure must be

LEFT: BLACK VULTURES
RIGHT: EASTERN PHOEBE BROODING NESTLINGS ON WINDOW SILL

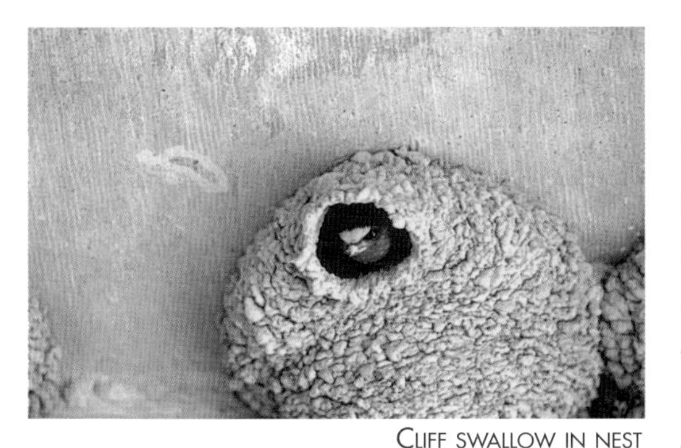

Cliff swallow in nest

removed, inspect it to determine if bats, chimney swifts, barn swallows, owls, raccoons, or other wildlife are inhabiting it. When possible, defer demolition or removal until the animals have left for the year. Many species, such as the chimney swifts, barn swallows, phoebes, and bats, will be present only during the nesting or breeding season. For any animals that might have taken up permanent residence, like raccoons, try to remove them before tearing down the structure.

Besides waiting to remove the structure at an appropriate time, consider creating some sort of a replacement for the animals. Bat houses and shelves for phoebes and swallows can be installed in the vicinity of the former structure. Check local nature stores or the Internet for information on purchasing these items.

## *Suggested Reading*

*Wood Projects for Illinois Wildlife (Homes and Feeders for Birds and Mammals).* 1997. Illinois Department of Natural Resources.

Cliff swallows nesting under barn eaves

*Chapter 8*

# *Backyards* and Other Small Tracts

*If you are one of Illinois' 2.5 million rural or suburban home-owners, you may have some neighbors that you've paid little attention to—all the wildlife, seen and unseen, who claim joint ownership of the area surrounding your dwelling.*

SPICEBUSH SWALLOWTAIL CATERPILLAR
PREVIOUS PAGE: LEFT, EASTERN BLUEBIRD FEEDING FLEDGLING; RIGHT, GREAT SPANGLED FRITILLARY

You may have noticed squirrels, cardinals, finches, different species of sparrows, and an occasional frog or garter snake sharing your backyard environment. But numerous other animals can be attracted to a backyard (or schoolyard, church or temple parcel, company or hospital property, or other similar tract) that has been designed to accommodate wildlife. An amazing variety of wildlife, hundreds of species in fact, can reside permanently or seasonally in small parcels of land near structures occupied by humans. Given the necessary habitat elements, these species will tolerate considerable human activity and readily coexist with us in our yards and other small tracts.

Wildlife aren't the only beneficiaries of backyard sanctuaries. People can get great aesthetic pleasure from both the plants and the animals. Nothing quite soothes the soul like sitting under a tall oak, gazing at a rainbow of garden color swaying in the afternoon breeze, or like enjoying breakfast at the kitchen table while watching cardinals, blue jays, and goldfinches eat theirs just beyond the window. Landscaping for wildlife also provides another benefit for the homeowner: increased property value. Shade trees, windbreaks, flowering shrubs, flower beds, and other plantings that improve wildlife habitat can boost a home's desirability and selling price.

BLUE JAYS

Schools, hospitals, houses of worship, private businesses, and public offices can benefit from providing improved outdoor settings for their students, patrons, or employees, who may use them for social gatherings, lunch breaks, and other activities. Large trees, bushes, and herbaceous plants decrease noise pollution and contribute to healthy air quality. Small spaces designed with wildlife in mind are good for public relations and good for the environment.

Landscaping for wildlife doesn't have to mean creating an environment uncomfortable for human habitation. A backyard, schoolyard, or business parcel can have "wild" areas with unmowed vegetation but still keep mowed paths, walkways, or other sections of lawn for the human visitors who prefer a mix of manicured and natural. And creating some living areas for wildlife neighbors will result in less outdoor maintenance time and cost.

This chapter provides a few ideas for taking wildlife needs into account on small parcels of land around homes, businesses, and community areas so that humans and native wildlife can continue to coexist in Illinois' changing landscape.

## *Backyard Habitat Issues for Wildlife*

Backyard habitats have increased dramatically in our state in the last hundred years. Illinois' population more than tripled from 3.7 million in 1890 to 12.4 million in 2000, and residential areas now cover 2.5% of our state. Many Illinoisans live in apartment and condominium complexes, but 55% of our state's housing units are single-family residences in urban and suburban settings, rural areas, and small towns. Each of these homes has at least some yard surrounding it. While many of these individual backyards are not large, their collective value is quite significant. Add to that all the properties of schools, religious communities, government, and business, and the potential benefit to wildlife is substantial.

COTTONTAIL RABBIT

One particularly important function of these habitats is to provide a network of stopovers for migratory birds. During the fall, hundreds of species of songbirds, waterfowl, shorebirds, raptors, and other birds travel from breeding grounds in the U.S. and Canada to wintering sites in the southern U.S., Central and South America, and the Caribbean. In the spring they return north. Birds need to stop along the way to feed and to rest. Because most of Illinois' broad expanses of agricultural lands offer little wildlife habitat for these species, backyards and other small tracts serve as migratory travel stops. The scattered nature of our rural and suburban small tracts augments the natural habitat found in the country. Small open spaces offer plant food, insects, water, and woody vegetation for perching and cover, all of which fill an important need.

Collectively, small tracts provide another important benefit. It may seem that the cottontails and cardinals, squirrels and toads of our residential areas are just a drop in the bucket as far as conserving Illinois wildlife. After all, our remaining natural habitat in rural Illinois harbors the largest numbers of these species. But in an era when scientists are seriously concerned about the loss of genetic diversity, hundreds of thousands of backyards and other tracts combined with larger rural acreages help provide significant genetic reserves, at least for the more common species.

Wildlife that live in or visit habitats used frequently by humans do face some difficulties that are less of a problem for wildlife in natural areas. Knowing about these threats will help backyard habitat planners avoid or minimize their effects.

- *Motor vehicles.* To minimize the threats posed by cars, trucks, and motorcycles, keep habitat improvements other than tree plantings as far as possible from roads.

Tufted titmouse

- *Lawn-care equipment.* Lawnmowers, weed trimmers, and cultivators can pose threats to unsuspecting wildlife. Before the activity begins, walk over the area to be covered to scare off animals and check for nests that can be avoided during yardwork.
- *Power lines.* Exposed wires in and around a building can threaten wildlife as well as humans. Check to make sure no exposed wires or uncovered electrical panels are present.
- *Dogs* and *cats.* Pet owners have to decide what they feel is most valuable—free-ranging pets, an abundance of wildlife, or some combination of both. If you have pets that frequent the backyard, remember that they are predators. Place feeders and houses to minimize predation.
- *Pesticides.* While various pesticides are used on rural acreages and farms, application is often intensified and concentrated in and around homes. In fact, statistics show that landowners apply pesticides to their lawns and gardens at rates many times higher than farmers apply to their crop fields. This increased use is caused both by preconceived ideas of what makes an attractive yard (no dandelions, for example) and by a desire for a certain level of comfort (such as keeping mosquitoes and wasps at bay). But many pesticides cause direct and indirect harm to wildlife. Most threats are indirect—for example, insecticides reduce the amount of insects, which in turn reduces a critical food supply for wildlife. Numerous studies, however, have also documented direct injury and mortality to wildlife from applications of common yard herbicides and insecticides.

Sometimes homeowners counteract their own efforts to improve their backyard habitat without realizing it. One example would be a homeowner who applies insecticides to a vegetable garden in one corner of the yard, near a butterfly garden created in another corner. Several butterfly caterpillars feed on common garden plants, such as tomatoes, parsley, and corn. Many insecticides used to eliminate garden "pests" are nonselective—they kill all moth and butterfly caterpillars at the site, along with dozens of other types of insects, including beneficial predators like ladybugs and preying mantis. Beauties that would grace the flowers of the butterfly garden are eliminated before they ever reach adulthood.

GIANT SWALLOWTAILS

There are numerous alternatives to using chemical pesticides in lawns and gardens. One of the easiest ways to reduce large numbers of pests is to increase the variety of plants on your property. The more varied your plantings, the more varied the types of insects that will be present. Since different insects prey on each other, more diversity will offer a better local balance. Lawn "weeds" can be removed by hand or smothered with a temporary covering. There are also various organic methods to repel or eliminate insect pests. For more information, check books on organic gardening or contact a natural-landscaping organization.

## *How You Can Help Backyard Wildlife*

Illinois homeowners, school administrators, business owners, organization trustees, and other small-tract owners can help wildlife by doing three things:

- Creating new habitat on their property.
- Protecting existing habitat on their property.
- Working with neighbors and local officials to encourage neighborhood-wide habitat development and protection, or working with local businesses, schools, and places of worship to improve their lands for wildlife.

LEFT: NATIVE PLANTS USED FOR BACKYARD PLANTING
RIGHT: BLACK SWALLOWTAIL CATERPILLAR

American robins feeding nestlings

## *Management Considerations*

Three general categories of wildlife use backyards and other small areas frequented by humans. Wildlife with small home ranges, such as toads, squirrels, cardinals, house finches, and garter snakes, may spend their entire lives in a backyard habitat. Migratory species whose winter or summer requirements are met in a backyard or similar tract may use it seasonally for nesting or for winter cover or feeding. Examples include summer residents such as house wrens and robins and winter residents like pine siskins and white-throated sparrows. Occasional visitors include species that might stop briefly during spring or fall migration, like a blackburnian warbler or a cedar waxwing, or local permanent residents that periodically visit a backyard, such as an opossum or a downy woodpecker.

Two factors affect the types and numbers of species that use and visit your backyard or other small-tract habitat:

- The type and amount of cover and food you provide.
- Your backyard habitat's proximity and connectedness to other suitable habitats in the area.

The more diverse habitat you create on your property, the more species you will attract. But to some degree, the surrounding land uses will also play a role in the numbers and types of species you

Indigo bunting

Northern leopard frog

attract. And whether your property is physically connected by habitat corridors to adjacent or nearby quality habitat will also affect the amount and type of wildlife using your property.

As discussed in chapter 2, every species has a different set of habitat requirements that dictates its home range. The more space and habitat variety offered in an entire neighborhood or rural area, the more wildlife that entire region will support. If your yard or other small tract is located near a park or forest preserve, for example, you will probably attract a greater abundance and diversity of species than if it is located near a busy shopping center. Backyards, schoolyards, and business properties located in rural areas often have other suitable habitat within the immediate area, and thus have a good chance of attracting a wide variety of species.

Some species with limited mobility, like frogs, toads, chipmunks, and voles, won't be able to move from the nearby habitat to your property if they have to cross large expanses of land without cover and protection from cars, cats, mowers, and natural predators. So it is important to have not only suitable habitat in the surrounding landscape but also corridors for travel. The corridors don't have to be large, although the bigger the better. For some species, a row of shrubs or a strip of unmowed grass or flowers may be sufficient to allow dispersal from one habitat area to another. But a broader corridor will give more species an opportunity for safe movement.

You can't control the location of your property, but you may be able to influence adjacent or nearby land uses. Why not encourage neighboring homeowners or businesses to develop wildlife habitat plans for their properties? Or if they don't want to invest in a property-wide plan, maybe as a minimum they would be willing to create a corridor between your land and a nearby habitat area.

Many homeowners fear repercussions if they strike out individually and create a drastically different "look" from their neighbors. One simple solution for suburban residents is to form a neighborhood wildlife-habitat group. With several neighbors working together on a project, individual concerns about image can be alleviated. A recent trend in subdivisions is the formation of homeowners' associations. This is another avenue to pursue neighborhood-based wildlife habitat planning.

The flip side of this trend is the residential restrictions some subdivisions and homeowners' associations have implemented. To keep subdivisions looking uniform and tidy, homeowners are required to use pesticides and aren't allowed to plant "wild" gardens or do other alternative landscaping. In this situation, any idea to

White-crowned sparrow

WILDLIFE CAN ADD MUCH TO A PEACEFUL URBAN NEIGHBORHOOD.

improve individual or subdivision habitats may be poorly received. Try showing photos of other subdivisions that have considered wildlife concerns, or take association committee members on a field trip to see how other neighborhoods have provided for both wildlife and human residents.

## *Wildlife Landscaping Needn't Be Wild*

Whatever decisions you make or restrictions you face, don't let wildlife improvements be a substitute for maintaining an attractive yard. Just allowing a yard to grow up without some sort of a scheme may look like you are testing the nuisance laws. Plantings for wildlife can be varied and extremely pleasing to the eye and can even include a variety of "weeds" that have wildlife value. Make sure your project plan provides the elements needed by wildlife and also incorporates contemporary aesthetics, especially in suburban settings.

Rural homeowners usually have more control over what they do in their backyards and don't face the restrictions encountered by their suburban counterparts. However, if property is surrounded by agricultural lands, the owner may need to consider the effects of a habitat project on adjacent crop fields. Discussing with the farmer (if someone else owns or farms the property) the location and type of plantings you'll be doing will help avert problems.

Rural homeowners can also work to provide better habitat by undertaking practices on any adjacent farm ground that they own or by encouraging other owners to do activities that enhance the overall area for wildlife. These might include placing any fallow areas or grassy field borders adjacent to the backyard instead of elsewhere on the farm; creating corridors between the backyard habitat and nearby woods, grasslands, or wetlands; leaving a few rows of crops unharvested adjacent to the backyard; and being particularly careful with pesticide use in the vicinity of the yard. Also, if the crop field is tilled each year, encourage minimum tillage on the field, or at least in a strip near the backyard.

Schools and churches might use a habitat project as a demonstration site and actively encourage nearby property owners to incorporate wildlife considerations as well. Habitat projects can often reduce the maintenance time and costs required to sustain a visually acceptable property. This fact alone may sell the idea of habitat development to an organization. Habitat projects also offer a focus for students or members of a religious or civic group.

## *Richard and Susan Day*

Nesting kestrels and bluebirds, ruby-throated hummingbirds flitting among summer blooming prairie plants, and summer tanagers feeding at one of many bird feeders are just a few of the highlights of Richard and Susan Day's backyard habitat. Their property's bird checklist features 172 species, with 60 of those nesting.

Though the yard has long been colorful, it hasn't always had this diversity of bird life. The Days live in Richard's grandparents' house, located near the southern Illinois town of Alma. His grandparents had an "old-fashioned" daffodil farm. Once the plants bloomed, the flowers were cut and shipped by train. The daffodils are now gone, replaced by prairie, a wetland, a few crops, and a yard. For a while the Days' yard was "typical," with a few trees and lawn. When they married in 1990, Susan learned about landscaping for wildlife and asked Richard, "Do you care if we plant a few things?" The answer was "No, do whatever you want." Susan took the message to heart, and over the years their "few things" have turned out to be more than 200 bushes and trees and twelve flower beds.

## *Developing a Plan*

Chapter 9, on planning, provides the basics of preparing a plan for habitat improvements and protection on your property. Be sure to read it once you're ready to create your plan.

Considering the human element is especially important when planning backyard habitats. Before deciding what portions of the yard will be devoted to wildlife, designate any areas of human use on your planning map, such as a volleyball court, barbecue area, clothesline, or vegetable or herb garden. The locations of these human activities may affect where you place various plantings, like trees that will become tall or flowers that will attract a lot of bees.

Conversely, it is important to think about how the human element may affect wildlife. For example, you may not want to place certain types of nest boxes near an area that gets regular use by humans, because wildlife parents will be disturbed while they rear their young. Or you might want to place hummingbird plants away from the street, where the birds could dart into traffic. And consider the effects on wildlife not only from your own family but also your

While the couple jokes that once they started buying plants they couldn't stop, each plant was carefully selected for its avian benefits. As more bushes and flowers were added, more birds came. The backyard landscape plan was carefully researched to provide the necessary habitats for birds. The Days were interested in providing not only food, but shelter. They were fortunate that nursery owners would search for native stock for them; this way the Days could be assured the plants would survive in southern Illinois. Today, their three-acre yard is the envy of many and perhaps is the "ultimate backyard."

Providing wildlife habitat doesn't stop with their yard—the Days also have a five-acre prairie. Once again, they researched their plan, contacting the local Illinois Department of Natural Resources (IDNR) office. From IDNR they learned about the Illinois Acres for Wildlife program, in which landowners agree to set aside land for wildlife. They worked with a heritage biologist in planning, preparing the soil, and planting. To complete their homestead, they established an eight-acre shallow-water wetland, ten acres of crop, and sixteen acres of woodland.

When asked what hands-on lessons they have learned, Susan replies, "With calluses on our hands, we learned how to be flexible, adapt, and that a whole lot of laughter helps! We also learned about patience. We would call the heritage biologist in a panic when only weeds were coming up and were reassured to 'have patience.' We have also learned to say no. You can't do everything. Even though we would like to go to the nursery and say, 'I want one of everything,' you must have guidelines on what to plant. And when you come home and find a big bag of forbs and two tree planters are on your doorstep, you know the plants must be put in the ground immediately—that's hands-on!"

Every year during the first week in May, the Days volunteer at Stephen A. Forbes State Park during conservation field days for the fifth graders of Marion County. Here excited kids learn about backyard wildlife and their food, shelter, and water requirements. The Days find this very rewarding—they note that the future is in children's hands, and they believe they can have a significant influence on these young minds.

*Susan L. Post*

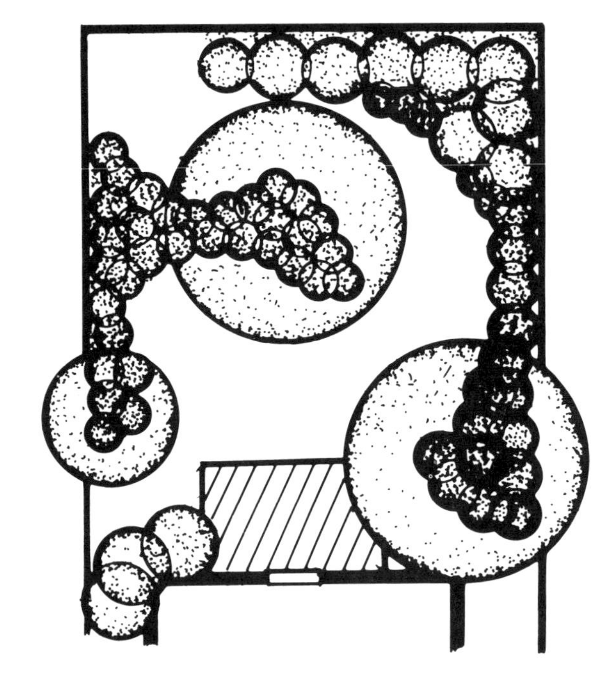

With just a little planning, diversity can be created, even in small backyards.

**Figure 8.1 Backyard Plans**

The best thing any homeowner can do is create diversity. Planting a variety of trees and shrubs will create a much better environment for wildlife and humans alike. Use half a dozen or more types of shrubs and at least three different tree species on a half-acre property—the more the better. And of course, even more diversity should be incorporated on a larger property. Look at the types of trees and shrubs

neighbors. For example, you wouldn't want to locate a nest box or garden adjacent to your neighbor's swimming pool if it will be busy all summer with activity and noise.

The following sections provide ideas on incorporating specific types of covers and features into your habitat plan.

## *Creating and Protecting Woody Cover*

Many Illinois yards and urban areas contain just a few species of non-native woody plants. For example, yards may have several yew bushes, a couple of shade trees, and nothing else. While yews may be used by house finches, cardinals, and a few other species for nesting, they provide little other wildlife value.

already growing in your neighborhood or on surrounding fields, and plant complementary native species to create diversity.

Shrubs come in myriad shapes and sizes, with flowering and seeding varieties. When selecting shrubs, choose plants that provide both good cover and a food source for wildlife. Plant taller species along property borders or behind grasses and flowers, and place shorter shrubs near homes, patios, and other human-use areas. Choose at least four different fruiting varieties to provide a varied food source for wildlife across the seasons. When possible, select native shrubs. While it may seem that exotic plants can't spread from a backyard, especially a suburban one, remember that the birds eating their fruits are transient and can deposit the seed remains of their meals miles away. Some of today's worst problems with exotic plants, such as Tartarian honeysuckle, in natural areas originated from plantings in backyards.

Trees offer dual benefits of providing both wildlife habitat and home-energy savings. Tall shade trees placed strategically on the south and west sides of a home or business protect the building from summer sun and can cut air-conditioning costs by 30%. Likewise, conifers such as cedars, pines, and spruces on the north and west sides of a building provide a barrier against winter winds, cutting heating costs up to 40%.

Deciduous shade trees can be planted near a building, but conifers should be planted farther away to maximize winter wind protection. A hundred feet between home and windbreak is ideal. Tall trees, both coniferous and deciduous, are good choices to define the boundaries of your property.

Table 8.1 recommends shrubs and trees for wildlife plantings. Also, see chapter 4 for details on establishing and maintaining woody cover.

CEDAR WAXWING FEEDING ON WINTERKING HAWTHORN FRUIT

**Table 8.1 Landscaping With Native Illinois Trees and Shrubs**

| Trees | | | | | | | |
|---|---|---|---|---|---|---|---|
| Common name | Scientific name | Full height (ft) | Regions where found* | Wildlife attracted | Food sources | Planting form | Native habitat (where species is best adapted) |
| Basswood, American | *Tilia americana* | 80 | All | Eastern chipmunks, songbirds | Seeds | Balled, seeds | Rich woods |
| Cherry, wild black | *Prunus serotina* | 75 | All | Songbirds, especially evening grosbeaks, robins, starlings, cedar waxwings; deer | Fruit | Balled, seedlings, seeds | Most conditions |
| Dogwood, flowering | *Cornus florida* | 50 | S, C | Songbirds, especially cardinals, evening grosbeaks, robins, wood thrushes, cedar waxwings; rabbits, squirrels, eastern chipmunks | Fruit | Seedlings, balled and burlapped | Rocky woods, wooded slopes |
| Gum, sour | *Nyssa sylvatica* | 85 | S | Robins, pileated woodpeckers, deer | Fruit, seeds | Balled, seedlings | Most conditions |
| Hackberry | *Celtis occidentalis L.* | 80 | All | Songbirds, mockingbirds, robins, small mammals, deer | Fruit | Balled, seedlings | Most conditions |
| Hickory, shagbark | *Carya ovata* | 80 | All | Songbirds, wood ducks, squirrels | Nuts, flowers | Balled, seedlings | Shaded woods, slopes |
| Hophornbeam (also ironwood) | *Ostrya virginiana* | 35 | All | Songbirds, small mammals | Seeds | Balled and burlapped, seeds | Upland woods, slopes near streams |
| Oak, bur | *Quercus macrocarpa* | 120 | All | Grackles, jays, brown thrashers, woodpeckers, small mammals, deer | Acorns | Balled, seedlings, seeds | Most conditions |
| Plum, wild (also American plum) | *Prunus americana* | 20 | All | Songbirds | Fruit | Seedlings | Woods, edges of streams, fencerows |
| Redbud (also eastern redbud, Judas tree) | *Cercis canadensis* | 35 | All | Songbirds | Seeds | Bare root, balled and burlapped, seeds and seedlings | Rich woods, ravines, fencerows |
| Redcedar, eastern | *Juniperus virginiana* | 35 | All | Songbirds, especially evening grosbeaks, cedar waxwings | Fruit | Balled, seeds | Dry areas |
| Sassafras | *Sassafras albidum* | 40 | All | Songbirds, spicebush swallowtails | Fruit | Seedlings, balled and burlapped | Roadsides, old fields, open woods |
| Witch-hazel | *Hamamelis virginiana* | 25 | N, C | Small mammals, deer | Seeds, bark | Seeds, seedlings | Moist or dry wood-lands |

| Shrubs | | | | | | | |
|---|---|---|---|---|---|---|---|
| Common name | Scientific name | Full height (ft) | Regions where found | Wildlife attracted | Food sources | Planting form | Native habitat (where species is best adapted) |
| Black chokecherry | *Aronia melanocarpa* | 12 | All | Songbirds, rabbits, squirrels, deer | Fruit | Plants, seeds, balled and burlapped | Moist, sandy woods |
| Dogwood, gray | *Cornus racemosa* | 10 | All | Songbirds, small mammals, deer | Fruit | Seeds, seedlings | Fencerows, road-sides, streambanks, prairie |

Shrubs *(continued)*

| Common name | Scientific name | Full height (ft) | Regions where found | Wildlife attracted | Food sources | Planting form | Native habitat (where species is best adapted) |
|---|---|---|---|---|---|---|---|
| Dogwood, red-twig (also red osier dogwood or poison dogwood) | *Cornus stolonifersa* | 10 | All | Songbirds, wood ducks, small mammals, deer | Fruit | Balled and burlapped | Marshes, moist fencerows, swamps, streambanks |
| Elderberry | *Sambucus canadensis* | 12 | All | Songbirds, small mammals, deer | Fruit | Seeds, seedlings, balled and burlapped | Moist soil, open woodlands |
| Leadplant | *Amorpha canescens* | 4 | All | Songbirds | | Seeds | Dry, sandy prairies, hills, open woods |
| Spicebush | *Lindera benzion* | 15 | All | Songbirds, wood thrushes, veeries | Fruit | Balled and burlapped | Rich, moist woods, streambanks |
| Sumac, fragrant (also aromatic sumac) | *Rhus aromatica* | 6 | All | Songbirds, small mammals | Fruit, bark | Balled and burlapped | Open woods, bluffs |
| Sumac, staghorn | *Rhus typhina* | 40 | N, C | Songbirds, waterfowl, mammals, deer | Fruit, bark | Seeds, seedlings | Dry soil, sandy ridges |

Excerpted from *Landscaping for Wildlife,* available from the Illinois Department of Natural Resources, which contains more information and additional species ideas.
*N: northern region; C: central region; S: southern region.

## Table 8.2 Butterfly Larval and Nectar Sources

**The following showy butterflies feed when they are larvae on the food plants listed at right**

| Butterfly species | Food plants for butterfly larvae |
|---|---|
| Black swallowtail *(Papilio polyxenes asterius)* | Carrots, parsley, dill *(Umbelliferae)* |
| Spicebush swallowtail *(Papilio troilus)* | Spicebush *(Lindera benzoin)*, sassafras *(Sassafras albidum)* |
| Tiger swallowtail *(Papilio glaucus)* | Wild cherry, birch, poplar, ash, apple trees, tulip tree |
| Monarch butterfly *(Danaus plexippus)* | Milkweeds *(Asclepias spp.)* |
| Question mark *(Polygonia interrogationis)* | Elm *(Ulmus spp.)*, hackberry *(Celtis spp.)* |
| Buckeye *(Precis lavinia)* | Plantain *(Plantago spp.)*, snapdragon *(Antirrhinum spp.)* |
| Red admiral *(Vanessa atalanta)* | Nettles *(Urtica spp., Boehmeria cylindrica)* |

**The following flowering plants are commonly used by various adult butterflies as nectar sources**

| Wildflower nectar sources | Shrub and tree nectar sources |
|---|---|
| Blazing stars *(Liatris spp.)* | Buckeye *(Aesculus spp.)* |
| Black-eyed Susan *(Rudbeckia hirta)* | Hawthorns *(Crataegus spp.)* |
| Butterfly weed *(Asclepias tuberosa)* | Fruit trees *(Prunus, Pyrus spp.)* |
| Coneflowers *(Echinacea spp.)* | New Jersey tea *(Ceanothus spp.)* |
| Ironweed *(Vernonia spp.)* | Lilac *(Syringa vulgaris)* |
| Joe-Pye weed *(Eupatorium spp.)* | Redbud *(Cercis canadensis)* |
| Milkweeds *(Asclepias spp.)* | Sumacs *(Rhus spp.)* |
| Phlox *(Phlox spp.)* | |
| New England aster *(Aster novae-angliae)* | |
| Wild bergamot *(Monarda fistulosa)* | |

ZEBRA SWALLOWTAIL

For more information on butterfly gardening and a longer list of plant types, see *Butterfly Gardens,* produced by the Illinois Department of Natural Resources.

Ruby-throated hummingbird

## *Creating and Protecting Herbaceous Habitat*

Grasses and broadleaf plants (forbs) play an important role in the backyard habitat plan. They add texture and color to the landscape and host a multitude of wild species, since ground-level cover is extremely important to nearly all wildlife. It is fine to designate areas of manicured lawn for human use, but be sure to also set aside spots where grasses and flowers will be allowed to grow to provide cover.

When thinking about grassy cover, consider eliminating your current lawn mix and planting a native prairie garden. Prairie grasses and flowers, many of which are warm-season plants, can provide an attractive alternative to non-native species such as bluegrass and fescue and are easy to maintain once established. Since they thrive on Illinois' hot summer weather, they won't turn brown in July and August like the non-native lawn grasses. Plant a balanced mix of grasses and forbs to provide variety. Consult chapter 3 for species selection. Consider prairie grasses and flowers that are smaller and less aggressive. For example, prairie dropseed could get the nod over big bluestem, and drooping coneflower would be a better choice than prairie dock.

Flower gardens that include non-native plants can also provide valuable cover and food for many wildlife species. However, as with woody plants, select native species as much as possible. And if you use non-natives, select those that do not invade natural areas.

Butterfly and hummingbird gardens have gained popularity in recent years. With the correct plants and an appropriate site, such a garden often attracts ruby-throated hummingbirds as well as dozens of butterfly species. Butterfly gardens should be located in sunny, sheltered areas protected from the wind. Butterflies are most active when warmed by the sun, so basking areas such as rocks and logs should be in full sunlight for most of the day. Include a couple of periodically moistened mud puddles as a source of water and minerals.

To keep butterflies visiting your garden all season, incorporate a variety of flowers that will bloom at different times. Also, be sure to plant flowers and grasses that will provide food for the caterpillar stage of butterflies as well as for the adults.

Table 8.2 (p. 181) details plants that attract butterflies and hummingbirds.

Monarch butterfly

## *Creating and Protecting Aquatic Habitat*

Adding some water to your backyard habitat can do wonders to attract more wildlife. Providing water can be as simple as setting up a birdbath or as involved as designing and excavating a small pond. Retention ponds, a common feature in many suburban and commercial areas, can also provide excellent habitat if properly managed.

Birdbaths have long been popular additions to Illinois yards. The most important aspect is to regularly fill the birdbath with fresh water. Often rain will do the work for you, but during dry periods, be sure to replenish the basin. This water source is of greatest importance to wildlife during extended dry periods.

When choosing a birdbath or water device, be sure it doesn't have a slick surface, so animals can perch easily on the sides and climb out from the basin. Also make sure it has a gradually sloping bottom and isn't too deep. Place it away from areas used by pets, especially potential hiding spots or elevated perch sites for cats. Remember, too, wildlife need water throughout the year. Extended winter periods below freezing are important times to provide a water source for wildlife. Consider installing a submersible heater in the birdbath.

Creating a small pond or pool in your backyard isn't as difficult as it may sound. Many nurseries sell plastic liners for creating ponds in porous soils. They also stock aquatic plants and other materials for creating a backyard pond. If your soil holds water, you won't need a plastic liner and you can dig out your own pond. Most small ponds will need to be replenished periodically with fresh water in the months when temperatures exceed 90°F. Include lots of plants in and around your pond, and consider placing logs and rockpiles in it or nearby.

Besides providing valuable wildlife habitat, a small wetland can be an attractive addition to a home or business. And it can provide excellent educational opportunities for schools. Seriously consider adding aquatic habitat to your property.

Retention ponds have become commonplace in many suburban areas. But many are little more than sterile bowls of water. With minor modifications, they could benefit a variety of wildlife. Consider planting some shrubs and trees near these ponds. Plant a portion of the edge with aquatic plants, and establish at least half of the banks with patches of tall herbaceous vegetation that will remain unmowed. Also consider building some rockpiles near the edge of the water.

Chapter 5 provides additional ideas on creating, improving, and managing aquatic habitats for wildlife.

CANADA GEESE

CARDINAL

## *Creating Winter Food Plots and Wildlife Feeding Areas*

RED-BELLIED WOODPECKER

Providing food for birds has significant value for certain species, especially during prolonged periods of deep snow or subzero temperatures. But most birds that dine at feeders also eat a range of other natural foods. Like people, most birds need a varied diet to obtain a balance of nutrients. To provide the best possible conditions for birds, provide a mix of food plots and bird feeders containing a variety of seeds.

Most native-flower gardens provide a range of seeds eaten by birds and other wildlife. Again, the more plant species you offer, the better variety for wildlife. And don't be quick to eliminate the "weeds" like goldenrod, foxtail, and asters that might pop up in your yard or garden. These plants usually have excellent food value. Also, if you have enough space, consider planting small food plots of corn, soybeans, sunflowers, or milo to supplement your natural food plots.

Nearly every bird-feeder design imaginable can be found on the market today; choosing the best one may seem a confusing task. If you have the space and the money, install three or four different types, such as a thistle feeder, a platform feeder, and a suet feeder. Feeders don't have to be commercially constructed to be useful. Placing seed on an old tree stump or similar structure also works fine. Platforms can also be built out of scrap wood. Some species prefer to feed on the ground, so spreading a little seed there is useful.

Try to shelter any bird feeder from winter winds. When possible, locate feeders next to shrubs, deciduous trees, or evergreens. This will not only protect the feeding birds from winter winds, it will give them nearby perching sites and provide some protective overhead cover from hawks looking for a feathered meal.

## Other Additions to Your Backyard Habitat

Most of the features discussed in chapter 7 can be incorporated into backyard habitats. Nest boxes are among the easiest and most visually acceptable of the practices. Many cavity nesters are adapted to human-dominated areas and will use nest boxes.

What about brushpiles or rockpiles, which neighbors might consider unsightly? Try hiding these in back corners of the property, with tall flowers planted around the perimeter. Dense shrubs can also hide a brushpile or rockpile. And wildlife will actually like it better when the pile is a bit hidden and has a food source nearby.

Creating backyard habitat can be a rewarding adventure. Numerous publications go beyond the scope of this chapter. Check with the Illinois Department of Natural Resources, a nearby botanical garden, or a local garden center or bookstore for additional materials.

## When Wildlife Wear Out Their Welcome

When you provide habitat and food for wildlife, you may get more than you bargained for. Species you didn't plan for may show up, literally at your door. And even desired species may overstep their welcome.

We enjoy the entertaining antics of a family of raccoons, the incredible ingenuity of a gray squirrel, the grace of a white-tailed deer. Watching a doe and her speckled fawn just outside the kitchen window on an early summer evening makes developing backyard habitat all seem worthwhile. Later, however, when deer browse prized garden plants or raccoons take up residence in the garage attic, our attitudes may change.

The homeowner who creates songbird habitat or a butterfly garden may later realize that other species may be attracted as well. Opossums may show up at night to feed on leftover birdseed. Or a red-tailed hawk may be attracted to the bird feeder—not for the seeds, since it is a carnivore, but for other birds as potential prey.

Yes, these are the facts of nature! Wildlife are hot-wired to survive. It can be difficult, perhaps impossible, to provide for only a few desired species without attracting others. This fact reflects the ecological adage that "everything is connected to everything else." It also speaks to the soundness of planning for more than just a few favorite species.

Occasional nuisances can be dealt with humanely and effectively. The resources listed at the end of this chapter can help you avoid problems before they start. Many are simple, common-sense approaches for keeping raccoons out of the house, reducing squirrels'

OPOSSUM

access to the bird feeder (unless you intend the feed for them, too), and protecting valuable landscape plantings from browsing.

The more we learn about wildlife, the easier it is to plan and enjoy a positive relationship with them. Inviting wildlife into your backyard can be a very important way to nurture this relationship. Watching squirrels from your dining room window provides year-round glimpses into their lives. From their mating chases twice a year, to their chattering communication system, to their scatter-hoarding of nuts in the fall, each seasonal observation provides a chance to better understand them.

Squirrels may challenge you to keep them out of the attic or the bird feeder, but remember, they have to make a living, too! You will find it helpful and enjoyable if you supplement your observations by reading about the life histories of a few of the animals you have attracted. Knowledge leads to understanding. And understanding can lead to tolerance and a peaceful coexistence.

## *Suggested Reading*

*A Field Guide to Your Own Back Yard.* J. H. Mitchell. 1999. Countryman Press, Woodstock, VT.

*Butterfly Gardening and Conservation.* D. Tylka. 1990. Urban Wildlife Series No. 2, Missouri Department of Natural Resources.

*Butterfly Gardens.* Illinois Department of Natural Resources.

*Controlling Damage Caused by White-Tailed Deer.* Illinois Department of Natural Resources.

*Keeping Wildlife Out of Your Home.* Illinois Department of Natural Resources.

*Landscaping for Backyard Wildlife.* D. Tylka. 1991. Urban Wildlife Series No. 3, Missouri Department of Natural Resources.

*Landscaping for Wildlife.* Illinois Department of Natural Resources.

*Living With Wildlife: Creating Wildlife Habitat No Matter Where You Live.* M. Leys and R. Leys. 2000. Krause Publications, Iola, WI.

*Native Plants in the Home Landscape.* K. G. Nowakowski. 2003. Circular No. 1381, College of Agricultural, Consumer and Environmental Sciences, University of Illinois.

*Nuisance Raccoons in Urban Settings.* Illinois Department of Natural Resources.

*Prairie Establishment and Landscaping.* W. E. McClain. 1997. Natural Heritage Technical Publication #2, Illinois Department of Natural Resources.

*The Backyard Feeder's Bible: The A-to-Z Guide to Feeders, Seed Mixes, Projects, and Treats.* S. Roth. 2000. Rodale Press, Emmaus, PA.

*Wild Neighbors: The Humane Approach to Living With Wildlife.* Edited by J. Hadidian, G. R. Hodge, and J. W. Grandy. 1997. Fulcrum Publishing, Golden, CO.

*Wood Projects for Illinois Wildlife (Homes and Feeders for Birds and Mammals).* 1997. Illinois Department of Natural Resources.

Fox squirrel

*Chapter 9*

# *Planning*

## Your Roadmap to Successful Habitat Management

*As with other activities, improving wildlife habitat requires planning. This is a crucial component because making landscape changes generally involves considerable time and effort and, once completed, the changes are fairly permanent. As they say, "If you don't aim, you won't hit the target." When you are investing a lot of time, toil, and money, it is important to develop goals and come at least close to reaching them.*

SAW-WHET OWL
PREVIOUS PAGE: LEFT, GREAT SPANGLED FRITILLARIES ON MILKWEED; RIGHT, CARDINAL

Developing plans for improving and protecting wildlife habitat need not be complicated; the level of effort is up to the individual landowner. But all management plans need certain elements to be successful. The following elements are recommended for any habitat project, whether in your backyard, a half-acre homesite, or on a 500-acre farm:

- Multiple copies of a map, aerial photo, or sketch of your property for an overall view of the landscape and a basis for outlining management strategies. Aerial photos can be especially helpful in examining the land uses of property surrounding yours.
- A written inventory of the existing natural features and the plants and animals inhabiting your property and the surrounding area.
- A copy of *The Natural Divisions of Illinois* (in the suggested reading list in chapter 1), to research presettlement plant communities, plus other relevant historical information for the property, such as past land uses.
- Written objectives for the property.
- A written plan of action, a schedule of activities, or the practices planned.
- An outline of a plan to monitor and maintain results.

## *Obtaining and Developing a Map*

To develop an accurate plan of where you are and what you intend to do, you need some sort of map. For a small property, simply sketch a map on paper. If your property is large, contact or visit the USDA Farm Service Agency (FSA) office in the county where your land is located (formerly the Agricultural Stabilization and Conservation Service, or ASCS, office). The agency usually has aerial photos of each section in the county, and for a nominal fee they can provide a copy that shows your land. You should also obtain a soils map for your property from the Natural Resources Conservation Service (NRCS), usually located in the same building as the FSA.

Many county courthouses have aerial photos that can be reproduced and purchased for a fee. Since these photos are often on a larger scale, they might be a better choice for a small property. Check with the superviser of assessment's office in your county.

Once you've outlined your property on the aerial photo or in a sketch, make a few photocopies for working copies in case you make mistakes or change your mind. Another option is to purchase from an office supply store clear Mylar overlays that can be marked on and erased. This strategy protects your original photo.

With the working copy in hand, walk over your land and mark the natural and constructed features on the map. If you're using an aerial photo, certain features will be apparent. Ask an FSA or tax accessor's employee to interpret features on the photo if you aren't familiar with using one. Also note any significant features you can on adjacent properties.

AERIAL VIEWS OFFER A LANDSCAPE-LEVEL PERSPECTIVE.

**Table 9.1 Features That Should Be Recorded on Your Map**

| | |
|---|---|
| **Natural features** | Soil types and slopes; cliffs, rocky outcrops; hills; caves; springs, seeps; large trees |
| **Existing habitats (with type, shape, and size)** | Forests, woodlots; fencerows, brushy thickets; grasslands, meadows; croplands, food plots; streams, rivers; ponds, marshes, swamps, lakes |
| **Wildlife use patterns** | Denning sites; rookeries, nesting areas; trails, travel lanes; hibernation sites |
| **Human-made elements** | Houses, buildings; gardens, orchards; roads, lanes, trails; grassed waterways, tile lines; livestock pens, pastures, watering tanks; utility lines, including buried cables and gaslines; erosion control structures; trash dumps; fences and wooden fenceposts; deer stands; nest boxes, feeders, waterers; brush-piles, rockpiles |

Also record any easements on your property that may be subject to future land changes. You would not want to create wildlife habitat only to have a utility company or governmental entity appear a few years later and eliminate the results of your hard work!

## *Written Inventory and Historical Records*

Once you have a basic map, you can begin looking at your property on a more intimate level. Whether it's your backyard or your back forty, you need to know what is there now and what was there in the past.

To keep organized, use a notebook and some sort of a pocketed folder or binder. You can divide records into "past" and "present" to keep material sorted.

EXISTING HABITATS SUCH AS THIS UPLAND MIXED-HARDWOOD FOREST ARE A PRIMARY CONSIDERATION IN PLANNING.

## *Inventory of Current Land Use and Features*

Start with an assessment of current land use, including those mentioned previously along with such activities as mowing and pesticide use. Once you have features and current land use recorded, you can begin to see areas where changes could be made to improve wildlife habitat.

Next determine what plants and animals currently live on or use your property. Some organizations will help with informal property surveys to determine the occurrence of animals and plants. But if you are making wildlife management a lifelong commitment, you can begin learning the species and do the survey yourself. Start by obtaining a few good nature field guides and walking over your property, identifying what you see. The suggested readings at the end of the chapter list selected field guides.

To supplement your field guide learning, consider joining a local conservation organization that takes regular field trips. This can be a quick and sociable way to learn many plants, birds, reptiles, amphibians, and invertebrates. Visit a local state park, interpretive center, botanical garden, or natural history museum that contains live or preserved specimens of native flora and fauna. Some parks and nature centers offer identification field trips led by experts who can provide a wealth of information.

Document the plants and animals that you know are on the property. If you want extra records for your folder, photograph plants and animals, or make a collection of dry plant specimens. These can also be useful for comparison and identification.

## *Historical Records*

Determine what plant communities could be growing in your area (if the existing ones are not native) and what animals would be expected in your region. For example, it would be pointless to plan habitat for ring-necked pheasants in Union County, because their usual range doesn't extend to southern Illinois. Or planting tupelo gum on a dry ridge in far northwestern Illinois would be ill advised; this species is at home in the wet forested bottomlands in southern Illinois. Determining historical plant and animal ranges and habitats can guide you in your efforts.

WILD TOM TURKEYS IN FULL DISPLAY

DICKCISSEL

One way to determine general community types for your part of the state is to obtain a copy of *The Natural Divisions of Illinois* (see "Suggested Reading" at the end of chapter 1). In 1973, botanist John Schwegman identified and mapped the different natural regions that exist in Illinois. He looked beyond geography to define the state's regions. For example, the cypress swamps in the southern part of the state are quite different from the glacial pothole wetlands in northeastern Illinois. Even though both are wetlands, each represents a different type of plant community. Schwegman used landforms, soils, geology, climate, and plant and animal distributions to define our state's regional diversity. When the project was complete, he identified 14 natural divisions, which he further divided into 33 subregions. While there is much crossover of plants and animals between divisions, each division has floral and faunal characteristics that are either unique or play an essential role in defining it. A great variety of Illinois plants and animals are restricted to one or a few natural divisions. Determining what natural division your property is located in can help you find out what types of plants, plant communities, and animals were and could be found there.

To delve in deeper and determine, to the extent possible, the original plant communities of your specific property, check soil maps from the Natural Resources Conservation Service (NRCS). You can

## *Illinois Natural Areas Inventory*

In the mid-1970s, a comprehensive project was undertaken to locate, describe, and catalog Illinois' remaining high-quality natural areas. The Illinois Natural Areas Inventory (INAI) documented sites containing remnant plant and animal communities that had been relatively undisturbed since European settlement of our state, habitats of threatened or endangered species, areas with outstanding geologic features, and other unique natural areas. They are the best remaining examples of our state's incredible natural diversity before settlement, a bit like living museums.

Many of these areas are located on private lands. While the land on the inventory is only about 1/100th of 1% of Illinois' acreage, it is possible that some readers of this book own or live near INAI sites. Management of certain natural areas requires special considerations beyond the scope of this book. If your property has been identified as having an INAI natural area or populations of endangered or threatened species, you should contact the Illinois Department of Natural Resources (IDNR) or the Illinois Nature Preserves Commission (INPC) to learn more about how to best manage the property and about incentive programs that may be available.

If you don't know whether your land contains a natural area but you think it might, send your name, address, phone number, and the location of the property to the IDNR or the INPC and they can determine that for you. If you have seen a threatened or endangered species on your property, you can help by sending that information to IDNR for inclusion on the natural heritage database, which tracks and monitors the status of threatened and endangered species in the state.

THIS HIGH-QUALITY PRAIRIE REMNANT GROWING NEXT TO A CORNFIELD INDICATES THE AREA WAS ORIGINALLY TALLGRASS PRAIRIE.

also check the Public Land Survey notes and plats for Illinois, found at the Illinois State Library and the Illinois State Archives in Springfield and at some university libraries around the state. These notes and plats are the actual documents that were recorded by the early land surveyors. Created in the early to mid-1800s, the documents provide an interesting insight into our historic landscape. The plats show what areas of the townships were prairie, forest, or wetland and denote such things as caves, springs, cliffs, and other natural features. The accompanying notes describe what was encountered along the section lines as they were laid out or surveyed. For example, the notes might describe a 24-inch-diameter white oak at the corner of a section. The notes can be difficult to read and interpret, so you may need the assistance of a knowledgeable librarian. The state archives and the state library are your best bet.

One way to incorporate into your plan the information you've gathered is to make a two-column page in your notebook. On the left side, record existing plant communities, with individual plants directly beneath the community heading, along with animals you've seen in those habitats. On the right side, list—as best as you can determine—what plant communities, species, and associated wildlife were originally on different areas of the property. You can use this to determine what woody and herbaceous plants you may want to introduce or eliminate.

One other historical aspect of your property should be researched if possible: the property's land-use history. This includes agricultural use, drainage, channelization, mining and logging activity, landfilling, and so on. Any knowledge of previous efforts to manage habitat for wildlife can also be useful.

Historical information on land use can be worthwhile in a variety of ways. For example, if you want to create a wetland and your property was previously tiled, you may be able to restore the wetland simply by plugging up or breaking the tile. Other records can be helpful in planning species selection and knowing when to plant trees, shrubs, or vegetation. For example, certain agricultural herbicides persist in the soil for a year or more and might require a delay in planting susceptible plants. Soil erosion concerns might dictate planting a cover crop with your woody or herbaceous vegetation to hold roots or seeds in place.

For a landowner whose property has been in family ownership for decades, developing a historical profile may be as simple as jotting down first-hand knowledge or interviewing relatives. Other owners with more recently acquired land will have to do extra digging.

Another way to determine previous land use is to talk to former owners or longtime neighbors. The FSA and NRCS can often provide valuable information. You can also check to see if a management plan was ever prepared by a state biologist. If so, it might hold clues to former land use. Examining historical aerial photos and topographic maps can also provide answers. Additionally, the stage of plant succession as well as species of plants growing on a site can give clues to past land use.

THE SUMAC AND GOLDENROD GROWING HERE INDICATE THIS AREA WAS A RECENTLY CULTIVATED FIELD.

## *Inventory of Surrounding Land Use*

The last step in your information collection is to survey the general area where your property is located to get an accurate perspective of the bigger picture. Look for these elements:

- Land uses on neighboring properties.
- Clues that neighbors are managing and protecting their property for plants and wildlife, such as an Acres For Wildlife sign or a posted designation as a Natural Heritage Landmark, land and water reserve, or nature preserve.
- Parks, public nature preserves, state or federal forests, conservation or fish and wildlife areas, and private preserves.

How you manage your land could have a positive or negative effect on nearby land management efforts for wildlife, and vice versa. At a minimum, try to consider the relationships between your land-management activities and those on adjacent properties. Contact other area landowners, private and public, and discuss ways you can complement each other's efforts and provide a mutual benefit to wildlife.

## *Determining and Recording Your Objectives*

Before you can set clear objectives and develop a management plan, you need to answer some basic questions.

* What potential does your property have? Will the soils, topography, water resources, and other realities support the project you have in mind?
- Which wildlife could benefit? Does it make sense to direct efforts toward improving habitat for a narrow or wide range of species?
- What are your capabilities? How much time, effort, and resources are you willing to commit to a project? What additional resources can you garner?

BUCK WHITE-TAILED DEER IN VELVET

- How can you evaluate your efforts? By measuring progress or lack of it, you can make management adjustments as you work through your plan.

Every parcel of land and every landowner is unique. Some owners want to re-create a natural area and manage for all wildlife, while others have a species of particular interest. An inventory of your property, your wildlife interests, and the information and ideas you have gained from chapters 3 through 8 should help you decide what to do. As you develop ideas, jot them down to form a sort of mission statement. This written record of your objectives will be a useful reference point and will be interesting to review in future years.

Many activities necessary to create, enhance, or maintain habitat are physically demanding and sometimes expensive. Determine what is realistic for you to accomplish given the cost and labor involved. Don't be too ambitious at first and plant more than you can take care of in any given year; you risk frustration and even temptation to abandon the project.

Numerous financial incentives are available for landowners with habitat projects, and if you can't perform the necessary labor or don't have proper equipment, other options are available. Hiring a high school or college student to do some of the labor can be an alternative, provided he or she is properly trained. Occasionally you can find volunteer help by contacting conservation organizations or IDNR. And private consultants can be hired to do site preparation, plantings, burnings, and other work.

Develop measurable goals, in part because it is much easier to stay motivated when you see results for your efforts. Develop both short-term and long-term goals to help you achieve your desired ends.

MOURNING DOVES

## *Developing a Plan of Action*

Once you've determined your objectives, you need to develop a schedule of how you'll meet them. This will be the guide to help you stay on track to obtain desired results in the shortest time possible. In your notebook, divide a sheet into four columns and label them as follows: activity or practice, location of the practice, scheduled date of completion, and actual date of completion. This will help you determine how much activity you have planned for any given season and will be useful later to evaluate results.

Since habitat establishment and management require many tasks that must be performed at specific times of the year, consider scheduling them in an appointment book. Another way to remind yourself of seasonal activities is to create a specific habitat-management calendar. First, record monthly information that will be used year after year, and make a master copy to use when creating calendars in future years. Then you can add activities specific to the current year. Tasks that you remind yourself of could be simple ones like cleaning out nest boxes or checking for a particular blooming plant, or more complicated activities, like preparing for a prescribed burn or planting trees and shrubs.

This book is written to give you general information on managing land for wildlife, but there is often no substitute for learning directly from people who have experience in habitat management. If possible, talk with a biologist from IDNR when you are in the planning stages of your habitat development or management project. IDNR personnel can help you determine if you are on the right path. And don't forget that other landowners who've actually done their own habitat projects can also provide a wealth of information.

Remember that your planning document should be flexible. It is perfectly fine to adjust your objectives and plan of action any time you wish. But to be successful, follow up with your project and evaluate the results. The final chapter discusses in detail the process of following up on your project.

## Suggested Reading


*Field Guide on Silk Moths in Illinois.* J. K. Bouseman. 2003. Manual 10, Illinois Natural History Survey.

*Field Guide to Amphibians and Reptiles of Illinois.* C. A. Phillips, R. A. Brandon, and E. O. Moll. 1999. Manual 8, Illinois Natural History Survey.

*Field Guide to Butterflies of Illinois.* J. K. Bouseman and J. G. Sternburg. 2001. Manual 9, Illinois Natural History Survey.

*Field Guide to Freshwater Mussels of the Midwest.* K. S. Cummings and C. A. Mayer. 1992. Manual 5, Illinois Natural History Survey.

*Landowner's Guide to Natural Resources Management Incentives.* Illinois Natural Resources Coordinating Council. 1997. Illinois Department of Natural Resources.

*Mammals of Illinois.* D. F. Hoffmeister. 1989. University of Illinois Press, Urbana.

*The Birds of Illinois.* H. D. Bohlen, with original paintings by W. Zimmerman. 1989. Indiana University Press, Bloomington.

*The Fishes of Illinois.* P. W. Smith. 2002. University of Illinois Press, Urbana.

*Waterfowl of Illinois Abbreviated Field Guide.* S. P. Havera. 1999. Manual 7, Illinois Natural History Survey.

RACCOON

*Chapter 10*

# *Follow-Up*

## The Long-Term Plan of Action

*The loss and degradation of wildlife habitat didn't occur overnight, and restoring it can't happen in a season or even a decade. But you will have taken an important step if you join the efforts of thousands of other landowners who manage their properties responsibly. With a decision to develop and protect high-quality habitat on your property, you will help ensure a future for our state's wildlife. Providing habitat should be viewed as a permanent endeavor, one to be integrated into everyday land-use decisions. This chapter discusses ways to keep on course with your planned management activities and to get optimum benefit from your efforts, along with the important topic of how to influence land management beyond your own property borders.*

EASTERN BLUEBIRD
PREVIOUS PAGE: LEFT, WHITE-TAILED DEER; RIGHT, COMMON WOOD NYMPH

## *What to Expect From Your Plan*

If you've recently started one or more habitat restoration projects, you may already be seeing some results. The shrubs or trees are leafing out, or prairie grasses are emerging. Maybe a wood duck is checking out the nest box at the pond, or a pair of bluebirds have already claimed their new nest site.

Take satisfaction in what is already happening. You may wonder, though, whether you'll live to see some of your projects reach their potential. If you've planted seedlings with the aim of creating a mature deciduous forest, it will be several decades before that habitat materializes. But in the meantime, you can watch wildlife respond to the changes as your trees grow and mature. This in itself will provide personal fulfillment. Creating other habitats, such as a wetland or a native warm-season prairie, produces quicker visible results, with some established vegetation and associated wildlife evident in the first few years. Building brushpiles or installing nest boxes can yield immediate results. However long it takes to see results for your efforts, you can receive satisfaction from knowing that you're doing the right thing for our natural world.

But what about the nest box that remains empty, or the trees that leafed out for a month and then died? Sometimes things do go awry. Plantings don't always establish as they're supposed to, and sometimes the wildlife you're trying to attract actually slow your progress by eating the vegetation you planted. This is why patience is so important.

Setbacks can happen, even for the most conscientious and well-intentioned landowner. Know your limitations when scheduling management activities and realize that you cannot control certain factors, such as the weather. Take on what you can realistically accomplish and expect that the weather will not always cooperate. Some projects may take a little longer or develop a bit differently than anticipated, but be steadfast.

Remember, you can learn a lot from mistakes or mishaps. Many failures with new plantings are the result of planting a species on an inappropriate site, not taking steps to minimize wildlife damage, or allowing too much weed competition. Following the guidelines offered throughout this book will go a long way toward avoiding problems and correcting mistakes.

Be sure to celebrate your successes, and take heart that every bit of habitat improvement provides some benefit to wildlife. Enjoy each step of the process, even if the end result isn't exactly what you expected. You're heading in the right direction!

Keep in mind the lesson from chapter 2. The complex characteristics of good wildlife habitat aren't created overnight. And remember, you may be doing more good than you can see. For example, if you have indeed created suitable nesting or winter cover, you might not be able to easily see or locate the wildlife using it!

Oak species direct-seeded into former pastureland

BELTED KINGFISHER

## *Reviewing Your Plan*

In medicine, follow-up care after surgery is nearly as important as the operation itself. The same holds true for "operations" performed on the land. Whether the surgical procedures are minor or major, some post-operative care and long-term maintenance will be necessary to ensure that your habitat objectives are achieved.

As a physician would do with a patient's medical chart, you should periodically review your management plan to check your progress and determine any changes that may be needed. Things change on the landscape and you'll need to make adjustments occasionally. You may have a new idea or want to incorporate new land into your management scheme. You should review your four-column management schedule (explained in "Developing a Plan of Action" on page 198). Did the work get done as planned? Or did you plan more than you could accomplish for the timeframe given? Maybe you'll need to make some adjustments for future planned projects.

A maintenance schedule is essential to your management plan. There are two types of maintenance: short-term or temporary, such as controlling weeds until tree seedlings become established, and long-term or permanent, such as burning or mowing a grassland every few years. Specific practices in your plan, such as grassland protection with delayed mowing, generally describe long-term maintenance activities. Short-term activities may use similar techniques, but they cease once a particular planting is successfully established. The maintenance schedule can be part of your four-column sheet or a separate list. If you include permanent maintenance activities on the four-column sheet, then under "planned completion date" you can mark, for example, "annually" and record the month the work should be done. Apply the same system for permanent activities done less often than every year.

The best way to keep track of necessary maintenance activities is to transfer them from your schedule or plan of action to a calendar or monthly planner at the beginning of each year.

## *Monitoring and Documenting Your Results*

Documenting the progress of your efforts can be a great motivator. Even if you have had some disappointments, looking back in two or three years at all you've done will propel you forward. Recording your successes and failures can also help you avoid repeating mistakes. And documentation will alert you to problematic patterns. For example, if all plantings of a particular species fail to survive two or three years in a row, this would show up in your notes and signal the need to reevaluate your plant choice for that site.

Recording how wildlife respond to your habitat improvements is also valuable for tracking your success. Nothing keeps a person going like seeing the tangible benefits of an activity.

Here are some suggestions for measuring the progress of your work, in addition to recording practice-completion dates on the schedule of activities, as was recommended in chapter 9. But don't stop with these; create your own documentation ideas, too.

- Keep detailed seasonal and yearly logs of work accomplished, including details like the amount of time it took to plant or construct, weather conditions before and after planting, animal damage to particular plants, and the survival or failure of various species.
- Record how long it takes plantings to reach various stages—for example, when shrub species first bear fruit, when prairie grasses first produce seedheads, how long it takes legumes to germinate in a cool-season planting, or how many inches a year your young oaks grow.
- Document how plants respond to various management regimes—for example, the effects of a prescribed burn in a native warm-season prairie grass planting, which site preparation (mowing, spraying, or tilling) yielded the best tree-growing conditions, or what species showed up after you disked a fallow crop field.
- Do a before-and-after inventory of part or all of your property. (This was also discussed in chapter 9.) Compare the plant and animal species present before your habitat management activities and after the first and subsequent years of a particular activity or planting. For example, you might record what species were present before and after developing a marsh, what wildlife species increased in number after planting a cool-season grass field, or what new species appeared after constructing a brushpile. Or you might record what exotic plants were killed by a prescribed burn or how spring wildflowers responded to a woodland burn.

Mayapples and spring beauties after a late-winter woodland burn

- Document progress with photos as well as notes. Photograph plantings, special features, and management activities to show changes. With woody or herbaceous plantings, take pictures of every stage of development. Good benchmark photos result from identifying "photo stations": so you can accurately compare your subject over several years, take the shot each time from precisely the same location with the same camera and lens.

## *Magnifying Your Efforts*

Your efforts on behalf of wildlife can have a positive impact far beyond your own property line, sometimes with only a little more work on your part. Here are some ways to influence land management beyond your own property:

- Enroll in the Illinois Acres For Wildlife program. Many new members first learn of the program through a sign or sticker posted at a friend's or neighbor's property. Even if you choose not to advertise your participation, enrollment helps wildlife management statewide. The Illinois Department of Natural Resources (IDNR) keeps track of the overall wildlife habitat acreage protected and developed in Illinois to help monitor the health of wildlife populations and communities. Why not contribute to this effort?
- Host an agency-organized field day at your property. IDNR biologists and local Soil and Water Conservation Districts are often looking for model properties where they can help other landowners learn about good wildlife habitat management. The agency will arrange the event; usually all you have to do is provide the site.
- Hold your own field day. Why not host a habitat party and invite your friends and neighbors out to see what you've accomplished. This can stimulate considerable interest and may lead to collaborative efforts. Another idea is to invite guests to the property before you've accomplished your goals and get them involved in the actual planting. The old barn-raising concept

SPRING FIELD DAY AFTER A LATE-WINTER BURN

could be applied to a large tree-planting party or prairie-forb planting event. Gathering a group to help you establish habitat not only provides assistance, it creates an opportunity for conservation education. To make the event extra special, have each participant plant and designate a "memorial" tree. When they visit you in the future they can check on its growth and health. Such connections to the land can help foster an interest in nature.

- If you're agreeable to publicity, recruit your local newspaper to do a story on your property. This can be a great way to spread the word about managing land for wildlife.
- Prepare a slide show or photo exhibit on your projects and advertise your availability to make presentations to schools, civic groups, and other organizations. In addition to showing what you've done, suggest ways the audience can get involved in habitat management.
- It has been said a thousand times, and each time it's true: children are the key to our future. Why not invite teachers or scout leaders to bring their kids out for a field trip? Or if you have children yourself, involve them in your work, then have them invite friends out for a habitat party, a prairie parade, or a tree-planting event. The possibilities are limited only by your imagination. You might encourage your child or children to get involved with your habitat improvement and management efforts as a project for school, scouting, or 4-H. Presentations about their project will educate their peers on habitat management.
- If you are a tenant farmer or you lease land for hunting, introduce your landlord to your habitat-management ideas. Lobby him or her to let you do some trial practices on the land.
- To help educate and motivate your employer or property-owning organizations you're connected with, offer a tour of your property or present any visual exhibit you create. Colleagues are often more receptive to a proposal if they can tap into the experience of someone who has successfully completed a similar project.
- Last but certainly not least is simply remembering to educate people one on one. Maybe a neighbor or friend has expressed an interest in nature or conservation. Take the person on a tour of your property and demonstrate the possibilities.

Caring for the environment and its wild inhabitants should be viewed not as an extracurricular activity but as integral to our daily lives. Whether you are planning to improve a small plot in your backyard or are gearing up to increase beneficial habitat on several hundred acres, you're heading in the right direction. Keep focused on the reason you're trying to improve your land, and you'll find that the rewards will exceed your expectations.

If all Illinois citizens adopted a pro-habitat perspective, just think where we'd be!

Carolina chickadee

## *Photo Credits**

**Richard Day/Daybreak Imagery:** title page 1, 6, 7, 9, 10, 13, 14 left, 15, 16 right, 19, 20, 26, 31 right, 32 left, 32 right, 40, 41, 46, 48, 50c, 50d, 50e, 53f, 55, 57, 64, 69, 70, 82, 86a, 86b, 86c, 86d, 86e, 86f, 95, 98, 101, 102a, 102e, 105, 110, 118, 132, 133a, 133b, 133c, 133d, 133e, 133f, 135 bottom left, 135 right, 137 left, 137 right, 139 right, 141, 142, 144, 145 top, 147 left, 147 right, 148, 151, 152, 153, 154 left, 157 right, 159, 160 inset, 160 right, 161, 162, 163, 164 left, 164 right, 165 left, 165 right, 166, 168 right, 169, 170, 171, 173 left, 173 right, 174 right, 179, 181, 182 left, 184 left, 184 right, 185, 188, 189, 193, 194, 197, 198, 200, 201, 202, 204, 207

**Todd Fink/Daybreak Imagery:** 39, 76, 85 right, 102h

Michael R. Jeffords/A Prairie Gallery: front cover, back cover, title page 2–3, 5, 8, 12, 16 left, 18 top, 18 bottom, 21, 22, 25, 27, 28, 29, 30, 31 left, 34, 37, 38, 42, 43, 45, 49, 50a, 50b, 50f, 50g, 51, 53b, 53c, 53d, 53e, 54, 62 left, 62 right, 71, 72, 73, 75, 77, 78, 79 left, 79 right, 81, 85 left, 86g, 86h, 88, 89, 91, 94, 96, 97, 99, 100, 102b, 102c, 102d, 102f, 102g, 103, 104, 106 left, 106 right, 107, 108 left, 108 right, 109, 114 left, 114 right, 115, 119 left, 119 right, 120, 121a, 121b, 121c, 121d, 121e, 122, 123, 124, 125, 126, 127, 128, 129, 130, 133g, 133h, 135 top left, 138, 139 left, 154 right, 155, 156, 157 left, 158 left, 167, 168 left, 172 right, 174 left, 175, 182 right, 183, 187, 190, 191, 195, 199

**Christopher A. Phillips:** 117 right

**Susan L. Post:** 53a, 116, 117 left

**Robert J. Reber:** 14 right, 17, 33, 36, 47, 52, 56, 59, 60, 65, 66, 67 right, 68, 74, 80, 83, 84, 93, 111, 113, 131, 134, 136, 140, 143, 145 bottom, 146, 149, 158 right, 160 left, 172 left, 176, 192, 196, 203, 205

**Ryan T. Reber:** 67 left

**David Riecks:** 206

**Mark Shanks:** 24

*Where photos are identified by letter, the lettering proceeds from left to right or from top to bottom on the page.

## *Illustration credits*

**Lydia Moore Hart:** 111 top, 122 bottom

**Lynn Hawkinson Smith:** 35, 51, 58-59, 90, 178

# Index

acorn planters, 89
aerial photos
acquiring, 191
agricultural-land management
contour buffer strips, 141
cool-season grasses for, 145
cropland management, 138
crop rotation, 138
field borders, 143–144
filter strips, 143–144
food and cover plots, 147, 149
grassed waterways, 142–143
hayland and pasture management, 138–140. *See also* hayland management; pasture management;
orchard management, 141
perch sites, 150
residue, 145–146
rotational grazing, 139–140
tillage, 145–146
warm-season grasses for, 144, 145
warm- versus cold-season grasses, 139–140
for wildlife, 137–141
American featherfoil, 121
American lotus, 99, 121
American toad, 108
amphibians, reduced breeding habitat of, 106
aquatic habitat, 183. *See also* wetlands
categories of, 100
creating, 118–123. *See also* wetland creation
defining, 100
den trees, 117
frog ponds, 122
islands, 117
issues in Illinois, 101, 103–107
marshes, 114
negative influences on, 104–106
plant diversity in, 114
snags, 117
successional stage of, 115–117
swamps, 114
types of, 118
aquatic wildlife, 117
arrow arum, 114
arrowhead, 121

backyard habitat
attracting butterflies to, 181, 182
creating diversity in, 178
food plots and feeding areas in, 184
herbaceous, 182
homeowner concerns, 174
issues for wildlife in, 169–172
landscaping considerations, 168, 175–176
management of, 173–175
migratory birds, 170, 173
planning for, 177
resources for developing, 185–186
selecting shrubs for, 178–179, 180–181
selecting trees for, 179–181
and threats to wildlife, 170–171
wildlife nuisances in, 185–186
woody cover in, 178–181
badger, 62
bats, 9, 86
beaver, 117
beaver lodge, 116
biodiversity
defined, 29
loss of, 20–21
biotic community
defined, 26
disturbances in, 30–31
birdbaths, 183
bird feeders, 184
birds
American goldfinch, 132
American kestrel, 50, 85, 133, 144, 145, 160
American robin, 173
Baltimore oriole, 9
barn owl, 19, 52, 153
belted kingfisher, 204
black vulture, 164
blue jay, 78, 168
blue-winged teal, 118
bobolink, 55
Bobwhite quail, 41, 47, 50, 55, 59, 92, 137, 157
brown-headed cowbird, 20
Canada goose, 21, 102, 109, 120, 162, 183
cardinal, 38, 184, 189
Carolina chickadee, 207
Carolina wren, 153
cedar waxwing, 179
chestnut-sided warbler, 76
chimney swift, 86
cliff swallow, 165
common yellowthroat, 38
dickcissel, 53, 194
eastern bluebirds 50, 78, 86, 154, 167, 202
eastern meadowlark, 51, 141
eastern phoebe, 164
eastern screech owl, 86
field sparrow, 50, 55, 58
grasshopper sparrow, 55, 64
great blue heron, 26, 107
great egret, 102, 109, 124
great-horned owl, 9, 95
green-backed heron, 7, 102
Henslow's sparrow, 16, 33, 59
hooded warbler, 19
horned lark, 19, 133, 135
house wren, 41
hummingbird, 182
indigo bunting, 173
killdeer, 139
mourning dove, 147, 198
northern harrier, 58
osprey, 117
ovenbird, 19, 39, 78
pied-billed grebe, 102
pileated woodpecker, 76
pintail duck, 108
prairie chicken, 14, 27, 52
prothonotary warbler, 117
red-bellied woodpecker, 184
red-headed woodpecker, 86
red-tailed hawk, 9, 133, 144, 150, 151, 159
red-winged blackbird, 19, 20
ring-necked pheasant, 32, 41, 50, 59, 129, 133, 139, 142
ruby-throated hummingbird, 182
saw-whet owl, 190
scarlet tanager, 78
sedge wren, 33, 59
short-eared owl, 57
sora rail, 105
tufted titmouse, 170
turkey vulture, 86
upland sandpiper, 19, 52, 59
vesper sparrow, 55
Virginia rail, 163
white-crowned sparrow, 174
wild turkey, 42, 71, 79, 92, 98, 193
wood duck, 86, 101, 102, 117
woodpecker, 86, 184
bison, 47
black oak-cedar upland forest, 83
black oak savanna, 94
black swallowtail caterpillar, 172
bobcat, 71
bottomland forests, 127
characteristics of, 75
composition changes in, 74
brome, 142
brushpiles, 185
construction of, 157
location of, 155–156
as microhabitat, 155
natural, 155
buildings and structures, 164–165
removal of, 165
bumblebee, 135
burning
benefit to woodlands, 96
in conjunction with overseeding and inter-planting, 69
effects of, 57
guidance for a prescribed, 96–97
prescribed, 66–68. *See also* prescribed burning
proper time for woodland, 97
timing, 67-68
in wetlands, 126
butterflies
Baltimore checkerspot, 53
common wood nymph, 201
giant swallowtail, 171
great spangled fritillary, 43, 53, 167, 189
monarch, 25, 45, 133, 182
tiger swallowtail, 43
zebra swallowtail, 181

carrying capacity, 33
cattail marsh, 104
chipmunk, 31
clearcutting, negative effects of, 95, 124
companion crops, 146
connectivity
in croplands, 136–137
of wetlands, 108
conservation efforts
effect of land-use policy on, 17–18
Conservation Reserve Program (CRP), 142
conservation tillage, use of, 134. *See also* tillage
contour buffer strips, 141
cool-season grassland, 49, 60
corn, 138, 148
corridors, 39
cottontail rabbit, 39, 50, 144, 155, 169
cover plots
planting, 147, 149
seeding rates and rotations, 149
coyote, 48, 69, 129
cropland changes
cropping intensity, 132, 134
effect on wildlife habitat, 132, 135
types of crops, 131–132
cropland habitat
crop stubble, 134
defining, 130
important changes in, 131-132, 134–136. *See also* cropland changes
issues in Illinois, 130–136
planting and seeding in, 148
poisoning of wildlife in, 134–135
types of crops, 131–132
wildlife in, 133
weed control, 148
cropland wildlife, considerations for helping, 136–137
crop rotation, 138
crop stubble, 134

cutting, 93–94
clearcutting, 95
of firewood, 94–95
group cutting, 95
of lumber, 94–95
proper time for, 94, 95
selective harvest, 95
cypress-tupelo swamp, 115

deciduous woodlands
creating, 87
defined, 72
spacing of trees in, 88
deep soil savanna, 75
den trees
defined, 87
proper number per acre, 95
in wetlands, 117
disturbance
burning, 57, 80, 110
in croplands, 137
cutting, 80, 110
development, 80–81
effects of woodland grazing, 79–80
erosion, 110
grazing, 57, 110
importance of, 36
invasive animals, 113–114
invasive plants, 112–113
lack of, 54
livestock watering, 110
mowing and haying, 56, 80
pesticide use, 57
pollution, 111-112
prescribed burning of prairie, 54
sedimentation, 110–111
tillage, 57, 109
types of, 36
domestic animals, influence on ecosystem stability, 44
drawdowns, 125
drooping coneflower, 67
dry upland forest, 75

eastern fence lizard, 79, 158
ecosystems
biodiversity in, 29
defined, 26
dynamic processes of, 30–31
effects of snake destruction in, 29
examples of, 27–28
role of fire in, 74
ecotones, 34
edge habitat, animal species in, 38
elk, 68
erosion
hindering with companion crops, 146
minimizing, 81
prevention of, 142
reduction of, 111
and wetlands, 110
evergreen groves, 72
creating, 87
evergreens
spacing of, 88

field borders, 143-144
filter strips, 124, 143-144
field windbreaks, importance of, 144
firewood, cutting, 94–95
fish
carp, 113
johnny darter, 122
longeared sunfish, 111
flooding
control of, 125–126
trees that tolerate, 125
flowers. See plant life
foliar chlorosis, 88
food plot mixes, planting details, 148
Food Plot Plant Selection and Management Guide, 148
food plots
planting, 147, 149
seeding rates and rotations, 149
forbs, 60
forest development, 35–36
forest successional stages, 84
foxes
gray fox, 89
red fox, 133, 137, 147, 156
fox squirrel, 133, 161
fragmentation, 19–20
frogs
bullfrog, 100, 102, 119
cricket frog, 106
green treefrog, 72
northern leopard frog, 104, 174

garden spider, 45
glacial lake, 119
goldenrod, 196
grains, seeding rates and rotations, 149
grassed waterway, 142–143
grasses
allelopathic, 90
big bluestem, 142
brome, 142
cool-season, 60
fescue, 91, 142
Indian grass, 142
little bluestem, 18
orchard grass, 142
redtop, 142
reed canary grass, 113, 142
seeding rates and rotations, 149
switchgrass, 43–44, 142
warm-season prairie grass, 60, 139
grassland habitat
aggressive exotic plants in, 52, 54
birds, 51
bluegrass in, 54
cool-season grasses, 60
creating grassy cover, 59, 61–64
decline of pastures, 51
defining, 48
forbs, 60
grassy cover, 48
legumes, 60
maintaining native versus non-native grasslands, 61
management consideration, 55
native prairie grasslands, 53
need for grass diversity, 57
perennial plants for food plots, 60
plant combinations, 60
warm-season prairie grasses, 60
grassland management, common methods for, 64.
grassland protection, 64
with delayed mowing, 65–66
with light grazing, 68
with prescribed burning, 66–68
with tillage, 68
grassland seeding dates, for Illinois, 63
grassland wildlife, 50
grass pastures, 139–140
grass-woodland ecotone, 34
grazing, 68–69
disturbance of wetlands, 110
negative effects of, 57
proper time for, 68
rotational system of, 139–140
woodland, 79–80
green-tree management, 125

habitat
backyard. *See* backyard habitat
carrying capacity of, 33
cropland. *See* cropland habitat
definitions of, 31
grassland. *See* grassland habitat
home range, 31–32
wetlands. *See* wetlands; aquatic habitat
woodlands. *See* woodland habitat
habitat degradation
as cause of limiting factors for Illinois wildlife, 32
and wildlife populations, 10
habitat improvement, measuring, 205
habitat management
considering interspersion in, 38
importance of using native plants for proper, 43–44
landscape perspective, 55. *See also* landscape-level management
maintenance schedule, 204
of the patch, 55. *See also* patch-level management
problem plants, 44
reviewing your plan, 204
habitat management planning
determining and recording objectives, 197–198
developing a plan of action, 198–199
surrounding land use inventory, 197
using a written inventory, 192–193
using historical records, 192, 193–196
habitat patches, 78–79
adjacent habitats, 79
connectivity in, 79
habitat research
land-use history, 195–196
*The Natural Divisions of Illinois,* 190, 194
Natural Resources Conservation Service, 194
Public Land Survey notes and plats for Illinois, 195
USDA Farm Service Agency (FSA), 191
using a written inventory, 192–193
using historical records, 193–196
habitat restoration projects
documenting results of, 204–206
Illinois Acres for Wildlife program, 206
maintenance schedule for, 204
management plan elements, 190
monitoring results of, 204–206
reviewing your plan, 204
sharing the experience with others, 206–207
habitat types, in Illinois, 31
hawks, 9, 133, 144, 150, 151, 159
hay fields, decreased wildlife in, 134
hayland management
producing crops and benefiting wildlife, 139
standards for, 138
herbaceous habitat, 182
herbicides
damage caused by, 135
use of, 64
and wetland wildlife, 112
hickory trees, 82
enabling mature growth of, 37
home range
defined, 31
of great-horned owl, 32
of ring-necked pheasant, 32
of white-tailed deer, 32
hunting, regulation of, 41

Illinois
bird populations, 20
cropland habitat issues in, 130–136
farmland in, 130
forest loss in, 16
grassland in, 52
grassland seeding dates for, 63
habitat types, 31
land ownership in, 10
landscape information for, 14–15
loss of wetlands in, 103
native plant species, 60
*The Natural Divisions of Illinois,* 190, 194
natural regions of, 194
number of species in, 8
old growth forests in, 76
prairie habitat in, 50
prairie loss in, 16
Public Land Survey notes and plats for, 195
trees in, 180
upland wildlife in, 123
water quality in, 42
wetland loss in, 16, 103
wetlands restoration guide, 123
wetlands wildlife in, 123
wildlife populations in, 10
woodland habitat acreage in, 73
Illinois Acres for Wildlife program, 22, 206
Illinois Department of Natural Resources (IDNR), 112, 194
Acres for Wildlife program, 22, 206
contact information, 11
Illinois landscape
loss of biological diversity in, 18
pre-colonial, 18

Illinois mud turtle, 18, 19
Illinois Natural History Survey (INHS), contact information, 11
Illinois Spring Bird Count, 20
Illinois trees and shrubs
  landscaping with, 179–181
  native habitat of, 180–181
  wildlife attracted by, 180–181
Illinois wildlife, importance of private landowners to, 22
insecticides
  damage caused by, 134–135
  and wetland wildlife, 112
insect populations, and pesticide use, 41
insects
  black-winged damsel flies, 99
  bumblebee, 135
  cicada, 53
  mayfly, 121
  twelve-spot dragonfly, 31
interplanting, 69
  defined, 97
interspersion
  defined, 38
  importance of considering, 39
  lack of in Illinois, 76
  of wetlands, 104
inventories
  of current land use and features, 193
  *See* Illinois Natural Areas Inventory (INAI)
  using nature field guides, 193
islands, value of, 117

land management. *See also* landscape-level management; patch-level management consequences of introducing non-native vegetation, 43
  of white-tailed deer, 27
landowners, helping wetland wildlife, 106–107
land preservation
  and farm policy, 17
  goals of, 21–22
landscape-level management (croplands)
  adjacent habitats, 136–137
  connectivity, 136–137
landscape-level management (grasslands), 55
landscape-level management (wetlands)
  considering connectivity, 108
  considering patch size and shape, 108
  watershed management, 108
landscape-level management (woodlands), 78
land-use history, researching, 195–196
leaf litter, 85
legumes, 60
  seeding rates and rotations, 149
limiting factor, 32
litter (dead plant material), 59
livestock watering, and disturbance of wetlands, 110
lumber, cutting, 94–95

maintenance schedule, 204
maps
  obtaining and developing, 191–192
  soils map, 191
marshes, 101, 114
mast
  hard, 82
  producers of, 82
  soft, 82
  and timber harvest, 83
mesic upland forest, 75
milo (grain sorghum), 149
  planting details, 148
mink, 102
moist-soil management, 124–125
  floods that arise from, 125
  proper water levels, 125
moths, 85, 97
mowing
  damage caused by, 135
  effect on nesting, 56
  improper time for, 139
  and protecting grasslands, 65–66
  of roadsides, 145
  in woodland habitat, 80
muskrat, 31, 125

native prairie garden, 182
Natural Resources Conservation Service (NRCS), 191, 194
nest boxes, 185
  design and maintenance of, 161
  species using, 160-161
nest cavities, artificial, 161. *See also* nest boxes
nest islands and platforms, 162–163
  wooden, designs for, 163
non-point source pollutants, 17
no-till, benefits of, 146

oak regeneration, 95
oak trees, 82
  enabling mature growth of, 37
  savanna relict bur oak, 37
old-field grasslands, plants of, 49
old growth forests, 76, 77, 88
opossum, 173, 185
orchard management, 141
overseeding, 69

pasture management, 138
patch-level management (croplands)
  disturbance, 137
  plant-species diversity, 137
patch-level management (grasslands)
  criteria of, 56–59
  disturbance, 56–57
  plant-species diversity, 57
  structural components, 59
  successional stage of, 58
patch-level management (wetlands)
  criteria for, 109
  disturbances, 109. *See also* disturbance
  seeding versus planting, 114
  structural components, 117–118
  successional stage of, 115–117
patch-level management (woodland), 79–86
  dealing with disturbance, 79–81. *See also* disturbance
  knowing successional stages of woody habitat, 84–85
  mast production, 82
  plant-species diversity, 81, 83
  role of den trees, 85
  role of snags, 85
  structural components, 85
pen-raised birds, survival success of , 41–42
perches, 150, 159–160
perennial plants, 60
pesticides
  alternatives to, 172
  dangers to wildlife, 171
  effect of, 57
  harm for wetlands, 104
  and insect populations, 41
  use of, 57
Pheasants Forever, 112
phragmites, 113
pin oaks, 88
plant communities
  dynamic nature of, 34
  succession of, 34–35
planting
  methods of, 89
  preparing a site for, 90
  in sandy soil, 90
  versus seeding, 114
  seedling care, 90–91
plant life. *See also* grasses
  American featherfoil, 121
  American lotus, 99, 121
  arrow arum, 114
  arrowhead, 121
  bluebells, 17, 126
  drooping coneflower, 67
  goldenrod, 196
  mayapple, 205
  milkweed, 189
  purple loosestrife, 112
  purple milkweed, 43
  spring beauties, 205
  sumac, 196
  turtlehead, 121
  wild blue iris, 121
  wild geranium, 80
  Winterking hawthorn, 179
  yellow lady's slipper, 97
plant populations, understanding, 34
plant species, native to Illinois, 60. *See also* plant life
pocket gopher, 53
pollution
  and wetland wildlife, 111–112
  of wetlands, 16
ponds
  in backyard habitats, 183
  nest islands and platforms in, 162–163
prairie habitat, in Illinois, 50
prairie marsh, 103
prairies, 49
  dolomite, 53
  dry to mesic, 53
  gravel, 53
  hill, 54
  Illinois' loss of, 16
  loess hill, 53
  sand, 53
  tallgrass, 51
  wet, 53
prairie stream, 111
prairie vole, 157
predators
  in Illinois, 44
  protecting nest boxes from, 161
Public Land Survey notes and plats
  accessing, 195
  for Illinois, 195

raccoon, 199
raptors, using artificial perches, 159
red-eared slider, 102
redtop, 142
reed canary grass, 113, 142
reforestation, shortcomings of, 74
residue, benefit to wildlife, 145
retention ponds, 183
riparian buffers, 124
river otter, transplanting, 42
roadsides
  cool-season grasses for, 145
  good grasses for, 144
  mowing of, 145
  Roadsides for Wildlife program, 145
  wildlife along, 144
rocky habitat, 158
rodenticides, damage caused by, 134–135
runoff, 104

salamander, 86
sand forest, 75
sand savanna, 75
savannas, 74
  creating, 87
  deep soil, 75
  defined, 72
  fire-dependency of, 96
  sand, 75
  spacing of, 88
sedge meadows, 101
sedimentation and wetland loss, 105
seeding, versus planting, 114
selective harvest, 95
selective thinning, 92–94
shrub borders
  creating, 87
  defined, 72
shrubs, spacing of, 88
shrub thickets
  creating, 87
  defined, 72
snags, 85
  creating, 93, 94
  defined, 87
  proper number per acre, 95
  in wetlands, 117
  wildlife dependent on, 86
snakes
  American broad-banded water snake, 126
  garter snake, 53, 133, 154
  Graham's crayfish snake, 117
  northern brown snake, 25
  prairie kingsnake, 50
Soil and Water Conservation Districts (SWCD), 89, 118

soil tests, 63
southern flatwoods, 75
special features
brushpiles, 155–157
nest boxes, 160–161
nest cavities, 161
nest islands and platforms, 162–163
old buildings and other structures, 164–165
perches, 159–160
rockpiles, 158
as structural components of habitats, 154
succession
assistance of, 37
defined, 34–35
disturbance during, 36
in Illinois, 35
large-scale, 35
stage of, 58, 116–117
sumac, 196
sunflowers, planting details, 148
swamps, 114
cypress-tupelo, 115
switchgrass, 43–44, 142

terraces, 141
tillage
in conjunction with overseeding and interplanting, 69
conservation, 134
damage caused by, 134
in grasslands, 68
managing, 145–146
minimizing, 124
no-till, 146
and wetlands, 109–110
timber harvest, 83, 95
timber management, 94–95
clearcutting, 95, 124
cutting timing, 95. *See also* cutting
group cutting, 95. *See also* cutting
oak regeneration, 95
selective harvest, 95
selective thinning of, 93
transplanting, success at, 42
trapping, regulation of, 41
tree-planting bar, 90
tree-planting machine, 89
trees
American beech, 88
black oak group, 82
evergreens, 88
flood-tolerant, 125–126
hickory, 82
of Illinois, 180
oak, 37, 82, 96
oak-hickory forest, 74
pin oaks, 88
white oak group, 82, 96
tubular tree shelters, 92
turtles
Illinois mud turtle, 18
painted turtle, 114
red-eared slider, 114

upland wildlife, in Illinois, 123
urban sprawl, and wetland loss, 105–106
USDA Farm Service Agency (FSA), 112, 119
U.S. Fish and Wildlife Service, 112

voles, 69, 144, 157

warm-season grasses, 60
warm-season grasslands
forbs in, 49
grass species in, 49
warm-season grassy cover, 61–62
water habitat. *See* aquatic habitat; wetlands
water levels
drawdown, 125
and food supply, 125
mimicking natural flooding, 125
winter, 125
water quality, in Illinois, 42
watershed, importance of protecting, 124
weed control, 91
for croplands, 148
wetland creation
attracting wildlife, 119
designing of wetlands, 120–121
determining objectives, 118–119
financial considerations of, 118
frog ponds, 122
introduction of fish, 122
physical factors of, 118
selecting plants for, 121–122
water depth requirements, 120–121
wetland management, 107–118. *See also* landscape-level management (wetlands); patch-level management (wetlands)
common activities of, 123
landscape-level, 108
patch-level, 109–118
planting methods, 127
wetland protection
filter strips and, 124
with green-tree management, 125–126
maintaining proper water levels, 125
with moist-soil management, 124–125
with prescribed burning, 126
riparian buffers and, 124
of watersheds, 124
wetlands. *See also* aquatic habitat
and clearcutting, 124
creating, 118–123. *See also* wetland creation
decreased quality of, 104
decreasing amount of, 105–106
defined, 100
fragmentation of, 104–105
Illinois' loss of, 16, 103
managing. *See* wetland management
nest islands and platforms on, 163
planting of, 127
pollution of, 16
problems presented by, 103
protecting. *See* wetland protection
reduced breeding habitat for amphibians in, 106
restoration guide, 123
runoff, 104
in southern Illinois, 101
successional stage of, 115–117
types of, 118
and urban sprawl, 105–106
watershed protection, 124
Wetlands Reserve Program, 142
wetland wildlife, 102
fish, 122
plant species, 121–122
plant-species diversity, 114–115
wheat, planting details, 148
white-tailed deer, 19, 29, 82, 91, 92–93, 99, 130, 139, 144, 197, 201
white oak, 96
wildlife
adaptation to buildings and structures, 164
ecological impact of, 9
factors affecting diversity of, 173–174
home range, 31–32
and the human element, 177–178
and Illinois trees and shrubs, 180–181
importance of private landowners to, 22, 106–107
landscaping for, 168, 175
life requirements of, 31
as nuisances, 185–186
providing food for, 184
roadsides and, 144, 145
wildlife feeding areas, 184
wildlife fencerows
creating, 87
defined, 72
wildlife habitat
fragmentation, 19–20
poor quality of current, 19
private versus government ownership of, 10
wildlife management
on agricultural land, 141–150
conservation organizations approach to, 40
levels of, 26
wildlife populations
factors that limit, 33
proper management of, 29–30
windbreaks, decrease of, 76
winter cover crops, 146
winter food plots, 184
woodland destruction
cutting, 93–94
girdling, 93
herbiciding, 93–94
woodland grazing, effects of, 79–80
woodland habitat. *See also* woody habitat
acreage in Illinois, 73
bottomland forest, 75
changing age of, 76–77
changing size and interspersion of, 76
changing species composition in, 74, 76
deep soil savanna, 75
dry upland forest, 75
issues in Illinois, 73–77
mesic upland forest, 75
sand forest, 75
sand savanna, 75
southern flatwoods, 75
woodland management, 78–86. *See also* landscape-level management (woodlands); patch-level management (woodlands)
allowing for variety, 81
interplanting in, 97
obtaining plants, 89
planting methods, 89
preparing planting sites, 90
prescribed burning, 96–97
protecting new plantings, 91–92
seedling care, 90–91
selection of plant species, 88
spacing of plants, 88–89
weed and grass control, 91
woodland restoration guide, 75
woodlands
biodiversity in, 83
destruction of, 93–94
riparian buffers, 124
soft mass plants and trees, 82
woodland wildlife, helping, 77
woody cover
in backyard habitats, 178–181
determining, 87
managing, 96–97
protecting. *See* woody cover protection
woody cover management, 92–97
woody cover protection
with prescribed burning, 96–97
with selective thinning, 92–94
with timber management, 94–95
woody fencerows, 76
woody habitat. *See also* woodland habitat; woody cover
animal damage of, 91–92
creating, 87–92. *See also* woodland management
defining, 72
high diversity of, 87
protecting new plantings, 91–92
successional stages of, 84
tubular tree shelters, 92
types of, 72
weed and grass control, 91